Chuck,

Best of luck
on your Glass Journey.

Joe
2008

JURAN
INSTITUTE'S
SIX SIGMA
Breakthrough and Beyond

Quality Performance
Breakthrough Methods

Foreword by Joseph M. Juran

Joseph A. De Feo
William W. Barnard

McGraw-Hill
New York   Chicago   San Francisco   Lisbon
London   Madrid   Mexico City   Milan   New Delhi
San Juan   Seoul   Singapore   Sydney   Toronto

The McGraw·Hill Companies

Library of Conrgress Cataloging-in-Publication Data

DeFeo, Joseph.
    Juran's six sigma : breakthrough and beyond / Joseph DeFeo, William Barnard.
      p. cm.
    Includes index.
    ISBN 0-07-142227-7
      1. Quality control—Statistical methods.   2. Production management—
Statistical methods.
    I. Barnard, william. II. Title.

    TS156.D3875 2004
    658.5'62—dc22

                                           203061833

    4 5 6 7 8 9 0   IBT/IBT   0 9 8 7

ISBN 0-07-142227-7

The sponsoring editor for this book was Kenneth McCombs and the production supervisor was Sherri Souffrance. It was set in Fairfield Medium by Patricia Wallenburg. Art director for the cover was Margaret Webster-Shapiro.

Printed and bound by Integrated Book Technology

McGraw-Hill books are available at special quantity discounts to use as premiums and sales promotions, or for use in corporate training programs. For more information, please write to the Director of Special Sales, McGraw-Hill Professional, Two Penn Plaza, New York, NY 10121-2298. Or contact your local bookstore.

 This book is printed on recycled, acid-free paper containing a minimum of 50 percent recycled, de-inked fiber.

CONTENTS

FOREWORD

Recent centuries have witnessed dramatic changes in managing for quality.

Until the industrial revolution (mid-eighteenth century), managing for quality was in good hands—the hands of craftsmen who performed all the tasks needed to make complete products—shoes, wagon wheels. That revolution created factories which then replaced many skilled craftsmen by semiskilled (or unskilled) task workers who performed only one or a few of the needed tasks. The resulting loss of the craftsmen's contribution to product quality then demanded expansion of product inspection and test.

The twentieth century then stimulated numerous dramatic innovations in managing for quality. These included:

- Creation of "independent" inspection departments
- Adoption of statistical methods for data collection and analysis
- Creation of "professional" job categories such as quality engineer or reliability engineer
- Realization that managing for quality warrants recognition as a new, scientific body of knowledge

In addition, the twentieth century witnessed the remarkable Japanese quality revolution which enabled them to become world quality leaders and an economic superpower. That same revolution then contributed to an enormous intensification of world competition in quality.

Meanwhile, the twentieth century witnessed a surge of growth in quality-related activities: professional societies, publication of books and papers, conferences, awards, consultants, and so on. At first, the publications focused on metrology, inspection, and statistical methods, but more recently the growing trend has been to books and papers on managing for quality.

The present book, by Joseph A. De Feo and William W. Barnard of Juran Institute, is a comprehensive contribution to this trend, with generous focus on the core processes by which managers manage—planning, control, and improvement. It is also wide-ranging—the case examples are drawn from multiple sectors of the economy. All in all, a welcome contribution to a vital human activity—managing for quality.

Joseph M. Juran

PREFACE

This book is about how to escape from "the predicament" that grips so many contemporary businesses and other organizations. As a leader in your organization, you may have read many books on improving and sustaining your business results. The authors don't want to waste your time on this book if it is not right for you. So please read this preface and decide if you want to continue reading.

If you can answer "No" to *all* of the following questions, then read no further. You are not in the predicament, and don't need to read this book (but you might pass it on to someone else who does).

1. Do you worry about how to satisfy all your various stake-holders—the shareholders, the board of directors, the customers, the employees, and regulators?
2. Are you doubtful about meeting your planned business results this year?
3. Do you feel like a victim of continual and unpredictable change?
4. Is your Six Sigma or continuous improvement initiative running out of steam?
5. Do you face unrelenting or mounting competition?
6. Do you worry about your ability to create customer satisfaction and loyalty?
7. Do some of your products or services become obsolete much faster than you ever imagined or planned?
8. Is your product development time cycle too slow to keep up or catch up with your competitors?
9. Do your processes produce waste, delays, defects, scrap, rework, and other avoidable excessive costs?

10. Does your organization need breakthrough improvements to become a worldclass outfit that offers world-class products and services?

11. Are your employees restive, seemingly uninvolved, indifferent, or hostile?

If you answered "No" to all these questions, then:

A. You may have misunderstood the questions.

B. You have addressed all of the issues raised by the questions (and we would like your phone number because we would like to talk with you).

C. You have your head in the sand, and probably won't be leading your organization much longer.

D. Your organization does not find itself in the same predicament that is reflected in the questions, and which confronts so many—if not most—contemporary organizations. (So please pass this book on to someone who needs it.)

If you answered "Yes" to any of the questions, then you may well be facing the predicament. If you are, we can offer you a way out.

Here's the predicament: How can you bring your organization to the point of answering "No" to these questions—and still meet your business goals, maintain a decent quality-of-work-life, and all without going financially, ethically, or emotionally bankrupt?

This book is written to help you loosen or entirely sever the grasp of the predicament, and protect you from allowing the predicament to catch up with you again. It is about understanding and executing the basics that drive sustainable breakthroughs in an organization.

INTRODUCTION

This book presents a typical predicament continuously facing organizations. It provides a strategy and a road map for dealing with the predicament and enabling you to achieve an organizational renewal to meet and survive the uncertain and rapidly changing future.

This book does not focus on technology (such as the details of IT systems or the latest engineering breakthroughs), although it assumes that your organization possesses contemporary hardware and software, and may be a technological leader. It is not, in that sense, a technological "how to" book. The book *does* focus on the managerial processes that can enable leaders to deal with the many dimensions of a human or social nature such as leadership, organization, management, culture, adaptability, planning, design, improvement, control, and other human activities. It also treats the inter-relationships of the human, social, and technological spheres of activity. We regard these human activities as crucial for surviving the future, because they embody the virtually infinite capacity of the human brain to distill meaning from apparent chaos, sense subtle shifts in the environment, and subsequently make rational decisions about what adaptive changes to implement.

The book is for both thinkers and practitioners—executives, students, managers, board members, and performers of any title, function, or level. It is oriented toward managing. Regretfully, we do not feel justified in claiming that the book presents a "unified field theory" of management, although it may represent elements of an eventual one. We also do not claim that we describe the only possible route to survival. There may well be alternate routes. What we do claim is that our message is conveyed to you from a succession of lessons learned from failures and successes over decades—centuries in some cases. We have drawn heavily from this knowledge.

Our conceptual foundation is the belief that all management consists of the essential functions of planning, controlling, and improving—the so-called "Juran Trilogy." The term "Juran Trilogy" was coined by the Juran Institute in 1986. It is based on observation and research conducted over many years by its founder, Dr. Joseph M. Juran. The trilogy refers to three managerial processes that must be understood and utilized to achieve sustainable high-quality results. For nearly five decades, the processes of the Juran Trilogy—planning, controlling, and improving—have been used as the basis for many quality systems, business improvement initiatives, and even Six Sigma. Dr. Juran believed, and so stated, that once these basic processes were carried out, an organization would have the means to sustain a competitive advantage and survive the future.

We have attempted in this book to better prepare managers to engage in all three of the managerial processes of the trilogy. They also need to know when to use each one, and why. Confusing them—employing one when another is more appropriate—can lead to unintended results.

A barrier to full realization of our prescriptions is a fairly typical managerial compensation system. Some of these policies cause honest, well-meaning, otherwise enlightened managers to do things "on the cheap" because every penny they save means a bigger bonus for them. However, simply cutting costs can, in the long-term, produce considerable customer dissatisfaction and an *increase* in overall costs, particularly costs of poorly performing processes (COP^3; waste). The phenomenon of cost cutting in one function that causes an increase in overall costs is called suboptimization. (Suboptimization is so deceptively appealing yet dangerous, that one utters the word aloud only in hushed tones.)

Consider this case of suboptimization. An early step in producing synthetic rubber is mixing. Typically, a separate factory is devoted to mixing the various ingredients, some of which are quite toxic and/or difficult to handle. Mixing must be done just so, to avoid undesirable damage to the rubber's chemical and physical properties with the passage of time. Mixing machines consume large quantities of energy. A plant manager of a mixing factory was under severe criticism from the Vice-President of Finance, and the President, for exceeding his energy budget.

To cut energy costs, the plant manager simply reduced the mixing time for each batch. Energy costs went down. The plant manager was out of the doghouse. But wait! Shortly thereafter, the next factories downstream in the manufacturing stream—extrusion, curing, molding, and cutting—began to experience many rejects. The scrap piled up, along with the associated costs of poorly performing processes. Soon a full-fledged emergency developed. "Why are we suddenly having so many defects? Why can't we work the rubber properly?" moaned the perplexed plant managers of the downstream plants.

You have already recognized the sub-optimization here. Partially mixed rubber can't do the job it's designed to do. Energy costs went down. Profits went down, too, because of the overall increase in the costs of waste and rework.

Cost cutting is easy in the short term, just slash and burn. What is not so easy is *reducing costs, while simultaneously increasing quality and customer satisfaction*. The latter is the subject of this book.

THE PREDICAMENT

Chapter 1 contains a review of the predicament faced by many, if not all contemporary organizations. The predicament is characterized by issues such as: how to sustain desired business results; chronic, accelerating, and unpredictable change; intense, unrelenting competition from home and abroad; product and service offerings that rapidly become obsolete; new product development times that are too slow to keep up with changing demands; non-competitive prices or minimal profits because of excessive costs—particularly costs of inefficient or wasted effort; irritating delays; chronic high levels of defects and errors with their associated scrap, rework, and dissatisfied customers; restive employees; unexpected challenges from continual new developments; and difficulties in recognizing and adapting to new realities. The list goes on. The reader undoubtedly can add to it.

We observe that an all-too-typical approach is to attempt specific improvements in only selected functions and levels.

The improvement efforts typically are not well-coordinated. Sometimes there are vague company-wide motivational campaigns with slogans and banners. One thrust competes with another for resources and the employee's time. Little if any time is allotted for these activities in the business plan. Employees are expected to perform them in addition to their "regular" job. Consequently, many employees simply don't have time to participate in performance-improving activities, and the activities falter or are ignored.

A Way Out of the Predicament

The piecemeal approach described previously may well result in some isolated and temporary improvements, but what is really needed is *sustained major improvement, organization-wide, built into the business plan*, what we call "sustainable performance breakthrough" (SPB).

Overall sustained improvement must involve many individual improvements, with each improvement involving all the functions, departments, and levels that contribute to each individual problem—and its solution. Organizations should not settle for simple improvements here and there, when survival requires major breakthrough across the board.

"Breakthrough," in this book, means deliberate *change*, a dynamic, decisive movement to new and unprecedented levels of performance.

"Performance" means *results*, as might be recorded on an executive scorecard:

- Shareholder value
- Profitability (ROI, ROA, ROS)
- Sales
- Market share
- Costs
- Customer/client satisfaction
- Customer/client loyalty
- Employee satisfaction

- Employee loyalty
- Cycle times
- Number of errors, defects
- Rework, redo
- Scrap
- Environmental citizenship
- Community citizenship

"Sustainable performance breakthrough" is not a one-time event. It is a continuous *process* that, once undertaken, is not only capable of rescuing an organization from its predicament, but also of preventing the predicament from reasserting itself. This book is offered as a prescription for doing just that.

The way out of the predicament starts with the actions of management, and ultimately involves all people in an organization.

THE BASIC TASKS OF MANAGEMENT: THE JURAN TRILOGY

According to the Juran Trilogy, managing consists of three basic processes:

- *Planning processes.* Processes to create innovative products, services, and processes.
- *Control processes.* Processes to prevent and/or correct unwanted "bad" change (to consistently maintain standards of performance).
- *Breakthrough processes.* Processes to create purposeful and unprecedented beneficial change (to improve upon current operations and to adapt to change—to prolong organizational life).

The core principles of the Juran Trilogy were published by Joseph M. Juran in his seminal book, *Managerial Breakthrough* (1964, revised 1995). The Juran Trilogy is Dr. Juran's conceptualization of the management process resulting from decades of analysis, synthesis, and reflection on scientific findings and lessons learned from observing how organizations operate. The trilogy principles apply to all types of organizations, all functions, and all levels. The book you are now reading is an explication of how to apply the Juran Trilogy in order to make your organization highly effective and efficient, and to assure its continuing survival. It's a guidebook for creating the organizational infrastructure and executing the activities that are required to achieve breakthrough performance.

THE JURAN TRILOGY®

Chapter 2 is an overview of the Trilogy, describing its interrelated elements and the purpose of each, the activities performed in each, and the benefits derived from each. As you examine each of the elements of the trilogy, you will notice that its activities center on *change*—creating desirable "good" change, and correcting or preventing undesirable "bad" change. Nothing could be more timely, or invaluable for the foreseeable future, because the survivors in a changing world will be the organizations who continually change themselves to keep up with new developments and avoid being left behind.

TRILOGY—PART ONE—PLANNING

Chapter 3 focuses on the first element of the trilogy: planning. It addresses the following questions: "Does my organization offer the products/services that customers really want? What should those products/services look like? How do we produce them?" The chapter describes essential planning functions—planning for quality—the design of innovative products, services, and processes. Numerous tools and techniques are reviewed, including Design for Six Sigma (DFSS). Note: In finance, planning is the equivalent of budgeting.

TRILOGY—PART TWO—CONTROL

Chapter 4 presents the second element of the trilogy: Control—a managerial technique carried out by the operating forces that assures that your products and processes perform to standard by preventing or correcting undesirable, unwanted, "bad" change in performance. In finance, one speaks of "financial control," which consists of measuring actual expenditures, comparing expenditures to budget, and taking action on overspending. In other operations, one speaks of "quality control," which consists of measuring actual performance, comparing actual to desired performance (the standard, the specification, etc.), and taking corrective action on bad differences. Control takes many forms, occurs in many places in an organization—at all levels—and utilizes many tools and techniques, some quite sophisticated and some quite simple.

TRILOGY—PART THREE—BREAKTHROUGH

Chapter 5 contains a discussion of the nature of breakthrough, which is the creation of sustained beneficial change to unprecedented levels of performance. We call this process "breakthrough," not "improvement," because it is not necessarily simple, easy, or quick. It often entails a good deal of detective work in order to discover root causes of poor performance, and it almost always requires penetrating the numerous sturdy organizational barriers that resist organizational change and protect the status quo. Rarely does breakthrough produce a sudden explosive burst of major overall organizational improvement (although localized sudden dramatic improvements could happen in a specific part of the organization as a consequence of, say, a Six Sigma improvement project). More often than not, a breakthrough, especially performance breakthrough, may take a while (months or years) to achieve because it is not, as we have stated, a one-time event. It is, rather, the aggregate result of many planned, coordinated, and meticulously executed individual improvement efforts in multiple functions and levels of the organization. A process breakthrough may happen quickly. A cultural breakthrough may take years.

BREAKTHROUGHS IN LEADERSHIP

Chapter 6 is the first of several chapters that focuses on a specific type of breakthrough. Each individual breakthrough type is essential to prolonging organizational vitality, although individually, none of the breakthrough types are sufficient. Performance breakthrough is the aggregate output of all of the individual types of breakthrough.

Chapter 6 addresses two questions:

1. "How do I set performance goals for my organization and activate the people in the organization to reach them?"

   Issues with leadership are found at all levels, not just at the top of the organization. The chapter reviews characteristics of effective leaders (i.e., those who get others to follow them), and the essential tasks of leaders: setting goals and strategies to reach the goals.

   In addition, the chapter points out some important differences that distinguish leadership from management. In general, leadership focuses on *change*.

2. "How do I best utilize the people and other resources in my organization? How should I best manage them?"

   The manager's role is one of administering the organization so that high standards are met, proper behavior is rewarded—or enforced, facilities and processes are maintained, employees are "motivated" and supported, and performance toward goals is monitored and "championed" to remove obstacles.. Management holds the organizational structure and operation in place. It also is continually planning ahead. In general, management focuses on *stability*.

BREAKTHROUGHS IN ORGANIZATION

Chapter 7 addresses the question: "How do I set up organizational structures and processes to reap the most effective and efficient performance toward goals?"

In addition to determining relationships and relative authority (the organization chart), and setting up the layout of workplaces, organization breakthrough concerns itself with the infrastructure required to foster performance that meets customer needs in an optimally economical manner. Examples are steering committees, councils of various sorts, work teams, project teams (such as improvement teams, design teams, and quality action teams), and self-directed work teams. Also included are various roles that managers take, such as champion, facilitator, supervisor, coach (yes, coach), etc.

BREAKTHROUGHS IN PERFORMANCE

Chapter 8 addresses the question: "How do I reduce or eliminate things that are wrong with my products or processes, and the associated customer dissatisfaction and high costs of poor quality (waste) that consume my bottom line?"

Breakthrough improvement addresses *quality* problems—failures to meet specific needs of specific customers, internal and external. (Different types of problems are addressed by the other types of breakthrough.)

Quality problems almost always boil down to just a few specific species of things that go wrong:

· Excessive number of defects.
· Excessive number of delays.
· Excessively long cycle times.
· Excessive costs of the resulting rework, scrap, late deliveries, dealing with dissatisfied customers, replacement of returned goods, loss of customers, loss of goodwill, etc.

Breakthrough improvement discovers root causes of the problems, devises changes to the "guilty" processes that remove or go around the causes, and installs new controls to prevent the return of the causes.

BREAKTHROUGHS IN CULTURE

Chapter 9 addresses the question: "How do I create a social climate that encourages organizational members to march eagerly together toward the organization's performance goals?"

Breakthroughs in culture focus on "soft" issues such as your organization's values, habits, beliefs, behavior standards, and social climate, all of which have a profound impact on performance. In addition, culture breakthroughs concern themselves with how to instill in all functions and levels the values and beliefs that guide organizational behavior and decision-making.

BREAKTHROUGHS IN ADAPTABILITY

Chapter 10 addresses two questions:

1. "How do I prepare my organization to respond quickly and effectively to unexpected change?

   Breakthroughs in adaptability (acting in concert with breakthroughs in organization) create structures and processes that sense changes/trends in the outside environment that are potentially promising or threatening to the organization, and evaluate the information from the environment and refer it to the appropriate persons for rapid adaptive action.

2. "How to I create and share the information that's needed to understand what is happening inside and outside my organization, so I know how things are going, and what we need to do different or differently?"

   It discusses how to relate—or match—information to your organization's performance and progress toward goals, so you can identify areas that require corrective action.

Numerous topics are discussed including—among many others—data and information chains; score cards; cost of poor quality studies; several types of surveys on: your organization's cultural characteristics, customers, clients, and the competition; and benchmarking for best practices.

STRATEGIC QUALITY PLANNING AND DEPLOYMENT

Chapter 11 covers strategic quality planning and deployment, a systematic process that sets organizational goals and makes decisions about what changes need to be made to reach the goals, what strategies to employ, who will do what, how performance toward goals will be measured, and who will take what action if performance toward goals is lagging. The outcome of strategic planning and deployment is an annual road map (a business plan) for reaching a given year's goals, and an organization whose members are unified in their understanding of what the goals are, and what contribution each function, department, and individual is expected to make to reach the goals. Strategic goals are formulated to exploit your organization's strengths and your competition's (if you have competition) weaknesses.

The chapter also addresses deployment, a systematic approach for converting strategic goals into action. Deployment also usually creates a *unified* organization, with each person understanding the strategic goals, embracing them, and aware of what specific contributions toward achieving the goals he or she is expected to make. Carried out properly, the entire organization marches together in the same direction, with most activity directly instrumental in reaching the strategic goals, and little activity wasted on extraneous or nonvalue-added "busyness."

A Strategy for Survival—Beyond Six Sigma

The book concludes with a relatively brief chapter that integrates the elements of the Trilogy into a coherent managerial strategy for you to employ in positioning your organization to survive the future. The final chapter contains a road map for change, and the nondelegable tasks that executives must carry out to sustain results. To smooth the way, we describe obstacles, cautions, and speed bumps to watch out for, and helpful hints in the form of "do's and don'ts."

ACKNOWLEDGMENTS

Producing a book is very much a team effort. In the case of this book, an inestimable debt is owed to the pioneering work and support of Joseph M. Juran. The authors' insights are based largely on his teachings, both in print and in person. We have made every effort to be faithful to them, and represent them accurately. In addition, his reputation has enabled us to carry out assignments and have experiences that likely would not have happened otherwise. The accumulated lessons learned are reflected in our text.

We also wish to recognize the Juran Institute staff and specifically Ignacio Babe Romero for his Chapter 7, "Breakthroughs in Organization" which he somehow produced while contemporaneously performing his major consulting and managerial responsibilities. Frank M. Tedesco and Randall Johns made major contributions to Chapter 3, "The Planning Processes," also while carrying out demanding consulting schedules.

We also thank Chris Bonner, Wayne Bombaci, and Robert Wilson who read drafts of chapters with a critical (merciless, even) eye, and made good suggestions for improvements. On many occasions, several folks at the Juran Institute home office, including Cindy De Carlo, Linda Ellrodt, Laura Sutherland, and Carole Wesolowski, enthusiastically lent a capable creative hand with research, creating visuals and other technical matters. Now that it is finished, each author wishes to thank the other for the needling and thoughtful suggestions each provided the other that pushed the book along.

Because Management is such an enormous subject, we fear that important points that deserve mentioning are missing not only from this acknowledgement but also from the text. Nevertheless, we have made choices, and tried to avoid writing about everything we know about everything *ad nauseam*. The responsibility for omissions or errors is ours alone.

Finally, we thank our respective families who shared some of the stress—and also the excitement—of making this book happen.

THE PREDICAMENT

In the fall of 2002, the Juran Institute was invited to visit Shanghai, China and speak on the subject of Six Sigma.

Six Sigma is a business improvement strategy that focuses on improving products, processes, and profits. It enables an organization to improve performance by eliminating deficient processes and defects in products and services.

We accepted this invitation with much excitement. Being able to travel to China and teach about a subject that was so important to us and every organization competing in this millennium was an opportunity that could not be passed up. We accepted the invitation and presented a one-day symposium on Six Sigma to over 400 Chinese executives. Much to our surprise, there were executives from many different types of organizations. Representatives from hospitals, hotels, manufacturing plants, financial services, and even an airline were in attendance. The questions they asked during the session left us wondering if something unusual was happening here in China and believing whatever it was would affect the United States and other countries in the near future. They asked many questions like how to deploy Six Sigma, what U.S. organizations were successful with it, and what were the failures. It made us wonder if it was "déjà vu." After all, a similar invitation was presented to Dr. Joseph Juran over 50 years ago by Japanese executives, and much to his surprise, they too were on to something. Twenty years after Dr. Juran accepted the invitation to speak to the Japanese executives on how to manage for quality, the Japanese manufacturers became the standard for excellent quality in many industries. The knowledge the Japanese executives gained from Dr. Juran's seminars was the start of the revolution in quality that took place in Japan, which later caused many U.S. and Western organizations to change the way they managed their organizations.

It may be too early to tell what the Chinese will do with the knowledge they have gained. Symptoms of another quality revolution are appearing in China and other low-cost producing nations such as Korea and Vietnam. They are making investments in training and education of employees on the principles of managing for quality leadership, business improvement techniques, and lean and world-class manufacturing. This, coupled with a strong commitment from the Chinese government to support these efforts, will likely lead to tough competition. This may even lead China to world dominance in quality at some period in the future.

The Japanese began in a similar fashion. In the 1950s they learned management techniques from Dr. Joseph M. Juran, Dr. Ed Deming, and others. Japan then spent the next 20 years improving the quality of their products and hence improving their business performance. Toyota still stands at the top of the list of manufacturers in having the fewest defects per automobile.

The Japanese revolution in quality forced many U.S. organizations to change the way they manage their businesses—if they did not, they would have gone out of business. Many did; some had not! Japan maintained this dominance for almost 15 years in a number of industries such as automobile manufacturing, high-tech electronics, and tool making. Many of these industries are still among the best-managed organizations in the world.

That was then, this is now. It is one thing to see Japan with around 100 million citizens become a global leader. It will be another to see what China can do with over 1 billion citizens. Armed with the same tools and techniques as the West and Japan, China would be a force to contend with around the globe.

That is what we observed from this one-day seminar. China is building a powerhouse built on low labor costs, high quality, and a supportive government.

China, in its strategy to be the global leader in the manufacture of the highest quality goods and services, is off and running, at a faster pace than the United States, Europe, or Japan could imagine. While the United States, Japan, and European organizations are dealing with an economic downturn, and in our opinion paying less attention to producing quality services and products, China is improving fast. In less time than the Japanese, China and perhaps a few other low-cost producing nations may emerge as a global quality leader.

The automobile manufacturer of Hyundai, a newcomer from Korea, is already in the top five of the JD Powers quality results. They have surpassed most the U.S. manufacturers with fewer defects per auto. They have only been in the United States for less than 10 years.

Training in improvement tools and technology is readily available. Bolstered by this training, it need not take much time to produce state-of-the-art, world-class products. If, and when, this occurs, there may be major upheaval in our economies and society. Many more jobs will be lost. Many more unnecessary bankruptcies and manufacturing plant closures will occur. Only this time it may take longer to catch up, unless our executives look at what they are doing to enable their organizations to remain competitive. Western executives need to increase their rate of improvement or they too will lose to the Chinese and other potential competitors which are improving at a faster rate.

Figure 1.1 shows that from 1950 to the late 1970s the United States had a competitive advantage in quality and performance. By the late 1970s Japanese organizations began their assault on the U.S. marketplace. The result had immediate impact on many U.S. organizations. Although the consumer benefited by having better products, the U.S. organizations lost market share and profitability at an increasing rate. The reason for this, in our estimation, is that the Japanese organizations improved at a greater rate than those in the United States. By

the mid-1980s, many United States organizations had started to
change the way they thought about quality and began to
increase their rate of improvement. Many areas were able to win
back markets lost to the Japanese. The United States has maintained much of this gain even until today. However, with low-
cost countries, such as Korea and China, producing the same or
greater level of quality products will this gain be maintained?
We believe unless the United States begins to increase its rate,
it may fall behind as it did to the Japanese in the early 1990s.
Organizations like Samsung Electronics, once considered a producer of shoddy electronic products, now leads the world marketplace in flat screen television production. Its vision is to be
the leading producer of high-tech electronics worldwide.
Looking ahead to the next decade (the dotted lines in gray in
Figure 1.1), the United States may actually fall behind the low-
cost producers. In just a short time, China will have learned the
same world-class manufacturing techniques the Japanese,
United States, and some European organizations have, and they
are applying them in many areas.

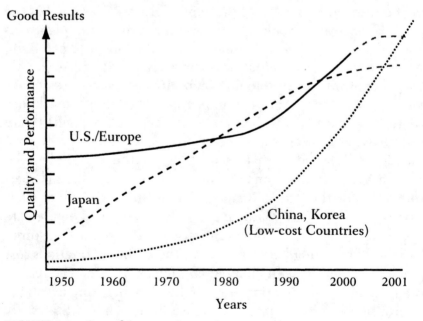

FIGURE 1.1   Rate of improvement.

Not all is bad for the United States and Europe. A number of U.S. and European organizations have improved significantly in the past decade using Six Sigma as the driver of quality improvement. For these organizations, they must continue for many years, not just one, two, or five years. The competitors from other countries, particularly low-cost producing nations, have much to gain. They will not stop improving. They need to improve their standard of living more than we do.

A concern we have is that many organizations that have adopted Six Sigma are missing some key elements that would, if they included them, enable them to sustain improvement into the future. If they do not adopt them, they will have difficulty achieving gains beyond their current efforts and possibly not at a rate fast enough to beat the competition. These organizations may not be prepared to meet competitive challenges in the future if they do not fix their initiatives today.

This book focuses on four major themes:

- The first describes the predicament that many organizations will face if they continue to improve at a lackluster rate.
- The second is about the current state of Six Sigma initiatives and how to assure yours is achieving the results you desired.
- The third is about the organizational breakthroughs needed to sustain Six Sigma results beyond a one- to three-year period.
- The fourth is a road map or set of activities that need to be carried out to enable organizations to move beyond their present performance—beyond Six Sigma.

We believe that we, at the Juran Institute, are in a good position to write this book. Our founder, Joseph M. Juran, was in Japan helping them lead the quality revolution in the 1950s. In the late 1970s and 1980s, we were working in the United States along with Deming, Crosby, and others leading the West with its revolution in quality. In the 1990s, we were there with the rebirth of Six Sigma with the likes of GE leading the Six Sigma way. Moreover, we are still here in the new millennium enabling organizations to sustain their efforts. At 98 years old, Dr. Juran is still here, although retired.

We want to provide this book to help executives understand what an organization can do to sustain a leadership position in results—profitability, quality, and customer satisfaction—over the long term. We want to help organizations avoid a pending disaster or be impacted by a Chinese quality revolution. We want to help them to avoid losing once again to better quality products and services. There is still time to avoid falling behind and not learning from the Japanese quality revolution.

This book is about "sustainability of performance"—how to sustain competitiveness and global leadership beyond a three- to five-year period.

In the late 1990s, about two-thirds of the Fortune 500 organizations had begun Six Sigma initiatives aimed at reducing costs and improving quality. Some like GE, Seagate, and Dow Chemical have had success. Some may only have success in the short term (less than 5 years). Others will abandon their efforts after only a pilot run at their initiative. One of the reasons for this is that many organizations do not truly benefit from their change initiatives, their cost reduction efforts, or they do not achieve the needed culture changes they hoped for. In a short time, their initiatives become just another "program of the year." Why? Because many of these organizations often did not understand the basic requirements needed to sustain their efforts. Sustaining breakthrough results with Six Sigma improvement initiatives requires fundamental knowledge about how breakthroughs occur and what an organization must do, learn, and maintain to sustain competitiveness.

So why is this an issue? Because now we are beginning to see a number of negative stories about Six Sigma that may cause many U.S. and European organizations to not improve at a fast enough rate. Many are not focused on long-term improvement, and some are beginning to abandon their latest effort to improve quality—the Six Sigma way—looking for a quick way to improve business performance. The Chinese executives desiring to improve at a faster pace should be an example and a wake up call for many Western nations and businesses to think about their strategies.

Samsung Electronics was once a shoddy manufacturer of low-tech electronic equipment. Today it is a global leader in cell phones, appliances, and flat screen TV monitors, to name a few. What did it do in the last decade? They improved quality through Six Sigma, focused on customer-driven design of new products, and managed assets effectively.

Yes, GE, Texas Instruments, Motorola, and even the Mayo Clinic were, and some are still, working hard at improving their results through Six Sigma and other quality and change management techniques. The question is whether they will sustain their efforts long enough to compete over the long term (greater than 10 years).

THE PREDICAMENT

Attaining quality leadership through Six Sigma improvement initiatives depends upon the length (number of years) of the effort and the rate of improvement (breakthroughs occurring at a rate to beat competitors). All too often improvement initiatives are abandoned because results do not occur fast enough or are not sustained long enough. This is a predicament that too many organizations have, and will continue to face.

What will it take for your organization, to achieve significant sustainable results that can become adaptable to the many external challenges it may face in the next decade? What can you do to improve your improvement results? Will you sit back and watch the Chinese, Korean, or a yet to be seen competitor reduce your market share and then your business results?

We set out to write this book to encourage organizations to not repeat the past. We are already seeing organizations that set out to improve performance through Six Sigma and then abandoned their efforts. They started out with the best intentions. The executives mandated "project by project" improvement. Many black belts were trained. Results were seen. Other issues took precedence. The executives backed off and are now moving onto other pressing matters, only to see the improvements not last. They then blame the Six Sigma techniques for not working for them.

How is it that these techniques can work for some and not others? Why is it that Motorola, which gave birth to the term "Six Sigma" as a methodology to improve quality and business performance, had to recently solicit GE to relearn how to do Six Sigma? The answers are actually simple. Some say these organizations did not maintain their efforts long enough. Others say they missed some key elements or the executives were not involved. Often these are correct reasons, but may not be the root causes.

The path of failure at implementing company-wide improvement or transformation is littered with organizations that have lost their way. We have identified six reasons for these failures, most of which are a result of macro events, not micro ones.

Here are our six reasons why organizations may not be sustaining financial performance and improvement over the long term and, therefore, must always try something new:

1. *Winning.* Becoming the market leader, the profit leader, the low cost producer, takes your eyes off the need to continuously improve. While an organization is winning back customers, sales increase, customer demand increases, and profits soar. The immediate success overshadows the need to look ahead at the macro-economic or societal issues that may cause a sudden change in the road ahead. Although no industry, like the airlines, can ever plan for something as shocking as the terrorist attack on September 11, they can do more to monitor trends, avoid quick declines due to competitors achieving breakthroughs, and win back market share.

   Ford Motor Company is a good example of this. For most of the 1990s, Ford improved its quality and began seeing impressive business results. As it succeeded, it began to dismantle the quality function and the initiatives to improve quality. This led to an increase in defects per vehicle. In 2000, Ford claimed to have had an astonishing 20 million warranty claims and $5 billion in warranty costs! What happened? Did Ford take its eye off the ball, was it enamored by winning, or did an executive make a bad strategic decision?

2. *A change in leadership.* All too often, an executive leading a transformation such as a Six Sigma effort retires, quits, gets fired, or dies, and the new leader tosses out the past leader's initiative, whether it worked or not. Often times this happens because the new executive wants to put his/her own stamp on an old initiative. These executives are well intentioned, but may not understand the impact of their changes when they make them. This in turn leads to discontent from the staff they are trying to encourage to get on board with a new program of change. This leads to the death of their program and dissatisfaction from the change agents.

3. *The organization gets tired of the current initiative and moves on too quickly*—not enabling a number of critical breakthroughs to occur that would sustain the efforts over the long term.

   A typical rallying cry, "I'll be glad when this Six Sigma initiative is over and we can get back to our real jobs" is heard. Heard enough and management creates a new slogan, a change in the techniques, a reduction in the training required to make the technique work, and soon you have a new program, a new name, and the same results—short-term, lackluster results.

4. *The company does not maintain an infrastructure to continuously train the work force over time.* The infrastructure in an organization should include the creation and maintenance of executive steering teams, champion support, training on tools and techniques, the goal deployment process, the review process, and the reward process that demonstrates the importance of the initiative to improve.

   In the early 1990s, Eastman Chemical Company won the U.S. Malcolm Baldrige Award. Within five years after the award, we had been invited to speak at the Chairman's Quality Day. We were surprised to be invited to this event. They had won the Baldrige Award, had increased shareholder value, and had done everything they needed to do to be successful. Yet we were asked to speak on how it could reinvigorate its quality initiative! When we asked why it needed to reinvigorate its initiative, it responded that "25 percent of the work force" had turned over (left Eastman) in five years and it had not trained its replacements with the tools and techniques of quality improvement in the same way. A common mistake was made. Eastman assumed new employees only needed to be trained on a portion of the training previous employees had received. When 20 percent of the work force that is not trained is mixed with the 80 percent that was trained, you get around 50 percent effectiveness (not mathematically correct, but you get the picture)! Sustain the infrastructure, the training, the goals, and the initiative, and you will sustain the results.

5. *A merger or an acquisition occurs*, forcing a transformation initiative such as Six Sigma to be postponed or derailed permanently.

   A well-planned merger often leads to a short-term focus on quality and an elimination of the Six Sigma initiative while the merger takes place. This often leads to customer dissatisfaction at a time when the projected revenue targets are expected but not achieved due to this dissatisfaction. Why? Because most mergers occur on paper, with many financial assessments but with too little knowledge about process performance and customer requirements that could prevent these failures from occurring. During the merger, costs are usually removed that immediately and often negatively affect customers, the supply chain, the quality, etc. Soon the customer loses confidence in the organization's ability to deliver. This leads to further cost cutting to make up for the loss in revenue, and a downward spiral occurs.

   Instead of maintaining the tools and techniques that could improve performance and meet customer requirements, they are abandoned for short-term needs. This often leads to a failed merger, failed results, and an abandonment of the Six Sigma effort.

6. *Macroeconomic events around the world impact our organizations greater and faster than ever before.* This leads to new initiatives that begin in our organizations to deal with these changes. This leads to new improvement initiatives instituted in our organizations every few years and a common statement of "here comes another one."

Almost every decade, we read about a new change process that takes hold in modern organizations. There were quality circles in the 1970s, quality improvement teams in the 1980s, re-engineering in the 1990s, and now Six Sigma. These initiatives usually take hold for a three- to five-year period. During the period, many industries move to a new level of performance. Improvements in profits, costs, customer satisfaction, and sometimes employee satisfaction occur. For each new period of renewal, organizations try to push the envelope of results higher and higher. Each period also begins by a clear proof of the need to change. Sometimes the proof is clear such as significant

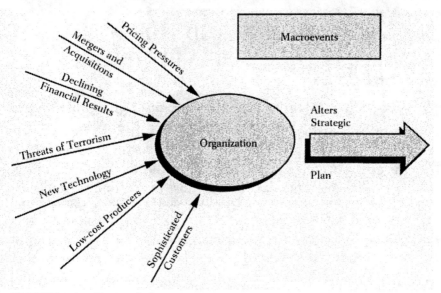

FIGURE 1.2   Forces of change.

losses due to poor customer demand, because of poor quality services provided to them. Other times the costs incurred due to poor performance reduces profits so greatly that an organization must act or go out of business. These new initiatives are necessary to compete. However, more can be done to prepare an organization for these phenomena. More knowledge is needed to know how to enable an organization to meet these events and continue improving at a rapid rate.

History has demonstrated that organizations often wait too long before initiating a change process to deal with the need to change. When the company decides to do something, it often does not know what to do differently than before, so it selects less than satisfactory techniques that often yield less than expected results. Many times, this could be avoided with better knowledge of the proper techniques to achieve sustainable change.

At times, organizations do not realize that change must occur as a normal process. The change process should be planned for, managed, and improved to enable the organization to remain competitive. In fact, most organizations react to change, rather than plan for it.

A LOOK AT THE PAST AND THE PROGRAMS THAT WERE CREATED AS A RESULT

We can look at history to see that organizations only improve when having a need to improve. In the 1950s, we saw the effects of World War II on organizations. After the war, U.S. organizations were required to produce products to meet multiple country demand. Japan and most of Europe were devastated during the war. Producing products as fast, and as good, as we could was the mandate. Many U.S. businesses met this challenge. As a result, Americans experienced a great deal of growth, improved economic lifestyle, and our products met customer requirements for this period—from 1950 to the early 1970s. In reality, the level of quality was dictated by the society and macroeconomics of the time. We are sure there were organizations that existed because they produced products that were superior to others and had produced positive business results. However, for the most part, the quality of products at any point in time seems to be dictated by the needs of our customers.

Because of the impact of the 1950s macroevents, new managerial techniques—"initiatives"—to meet customer requirements were created. At this time, most were "control-driven," as shown in Figure 1.3. These techniques enabled products to be produced at a record pace by maintaining a level of performance that by today's standard would be unacceptable. By the standards of its time, they were acceptable. Techniques such as MIL Standards 9858, and ISO Standardization and even the good manufacturing protocols (GMP) were becoming commonplace.

It was not until the 1970s that the United States began to understand that the quality of its products were deficient. In the late 1970s, the United States experienced an oil embargo that caused the price of gasoline to increase extraordinarily. This increase caused many citizens—customers—to look for alternative automobiles that operated at better miles per gal-

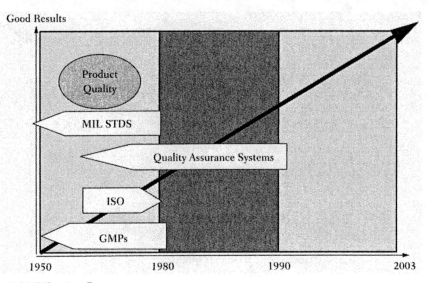

FIGURE 1.3   Response to events.

lon. They found them in the Japanese vehicles. Automobiles built smaller for Japan were making their way to the United States. Americans bought them and learned that not only were they more efficient than the U.S. cars, they were better. They broke down less often and ran longer. Their quality was better than the U.S. cars. This caused many Americans to not purchase U.S. automobiles. An alternative and new standard of quality was beginning to emerge. This led many organizations to react, and rethink their quality. A new generation with a focus on quality emerged. An American revolution began.

In the early 1980s, Ford claimed that it recalled more cars than it sold in the late 1970s, due to poor quality. Motorola made televisions sets that were noted for their defects. You may remember owning one of these televisions. We used to have to slap the top of the set periodically to make it work! Japanese products were making their way to the United States and they performed better and in many cases were less expensive to purchase. Japanese televisions worked. Their cars were better. A new standard of quality was defined.

U.S. organizations then had to react to this new level of quality. Why *react* versus not *anticipate* the challenge that they

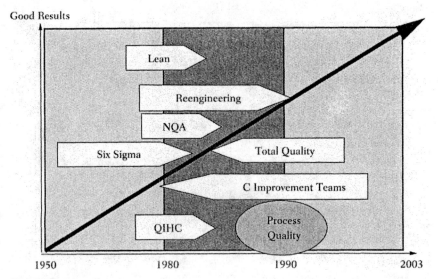

FIGURE 1.4   Response to events.

now faced? One hypothesis is that the American organizations were so busy producing products to meet demand that they failed to observe the progress of Japanese organizations. By the time they realized what was happening, it was too late. The response to this event was an extraordinary effort on the part of American organizations to improve quality. A revolution began and did not end until many U.S. organizations won back their lost business and beat the Japanese at their own game. The U.S. Malcolm Baldrige Award for Quality was created, Six Sigma at Motorola was underway, and lean manufacturing was being put into organizations. The focus on controlling products began shifting to improving the processes that created them. These improvement processes enabled efficiency and productivity to increase. The result was greater customer satisfaction, lower costs, and greater profitability.

However, just as we were improving, those macroevents kept haunting us, and in the early 2000s, U.S. businesses began to see a decline in customer demand due to a slowing economy, a terrorist act, financial wrongdoing in some large corporations, and a Firestone-Ford catastrophic event began a new concern about quality and performance in America. How could this happen again? This is the same period that GE, Dow Chemicals, American Express, and many other organizations began new Six Sigma initiatives. It happens because of the lack of a great enough rate of improvement, a lack of understanding about the drivers of customer demand (because of abandonment of quality surveys in the late 1990s), and a winning attitude once again, just like in the 1950s.

In the early 2000s, the response has been to focus on improving information quality and service quality. Techniques like Six Sigma, Design for Six Sigma (DFSS), Health Care Improvement (QIHC), and the Voice of the Customer (VOC) are the techniques being used.

There is a concern: this time it is not a countrywide revolution. It is a company-by-company revolution. Each company sees the impact of the macroevents independently of each other and is reacting. This, in our opinion, is creating a slower result overall on our society. This is happening at a time when the low-

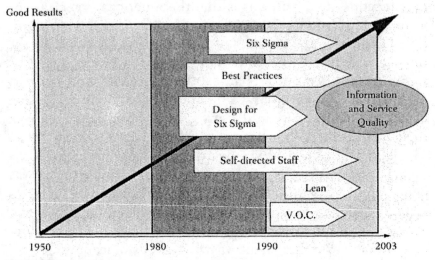

FIGURE 1.5   Response to events.

cost producers are improving their quality. The result may be that it takes us longer to recover and be the global leaders again. In the meantime, how many more jobs will be lost, shareholders dissatisfied, and companies forced into bankruptcies?

THE SOLUTION TO THE PREDICAMENT

We cannot avoid macroevents beyond our control. We can control our reaction to them and we can create organizations that will adapt to the changes needed. What is needed is a basic grounding in the principles and techniques that would enable the organizational breakthroughs that are necessary to occur in any organization to sustain long-term improvement. We, the authors, will provide management, executives, and each staff member with the knowledge of what to do differently to avoid these situations.

Managing for sustainable results requires basic knowledge of the managerial processes that lead to important breakthroughs that enable long-term improvement. An effective Six Sigma initiative must include not just the tools and techniques to solve problems, such as the DMAIC (define, measure, analyze, improve, and control) process, but also the organization-wide breakthroughs that must occur to sustain results over time.

Many books have been written about the means to achieve Six Sigma. Most describe the DMAIC process for improving processes, the DFSS or DMADV process for designing new services and products, and they identify many statistical tools to use to achieve Six Sigma levels of improvement. All of these are required, but alone do not bring about culture change or the ability to create long-term results for an organization.

We have set out in this book to describe not just the Six Sigma process, but also provide the means for an organization to achieve longer-term cultural breakthroughs and the means to go beyond Six Sigma. Doing this will prepare an organization to continuously improve at rates faster than their competitors. Our title "Six Sigma: Breakthrough and Beyond" is provided to give insight to the fact that quality improvement is never ending, that breakthroughs must occur annually, and that there is a means to get there. A road map exists.

As you proceed through each chapter, think about your organization and determine how well it understands the predicament and how well it is prepared to deal with it.

THE BASIC TASKS
OF MANAGEMENT:
THE JURAN TRILOGY

The way out of the predicament presented in this book has its origin in the work of quality pioneer and management practioner Joseph M. Juran. This book builds on management principles, as espoused in his lectures, books, and consulting. These principles are known as the "Juran Trilogy®." We believe that sustainable performance breakthrough is achievable by following the principles of the trilogy. Understanding the trilogy is the first step on the way. Actions by management follow.

THEORETICAL UNDERPINNINGS
OF THE TRILOGY: BASIC CONCEPTS
AND DEFINITIONS

We (the authors and the reader) need to establish a common language. Therefore, the following is a sequence of definitions and explanations that lead to the exposition of the trilogy.

- The *purpose of an organization* is to meet the needs of its clients or customers by supplying them with products and services at the lowest optimum cost (the combined costs to the supplier and customer). All organizational and individual performance is dedicated to that end (or should be).

- *Products* can be hard goods, services (work done for someone else), or information (data, software, etc.). Products are produced by means of processes.

- A *process* is a sequence of tasks or events that creates an output—a product: goods, services, or information. Processes include everything involved in each step: people, techniques, software, machines, tools, raw materials, facilities, procedures, management practices, etc.

- A *customer* is one who receives the output of a process or any step in the process. A *client* is a special kind of customer who pays for what is received. Customers do not necessarily pay, although they may. There are two basic types of customers: *external* (located outside of your organization) and *internal* (located within your organization). Each type of customer has needs that must be met if the organization is to serve its purpose. Usually, in a world-class organization, meeting the needs of internal customers is a prerequisite for being able to meet the needs of external customers. Management must devote its energy to meeting *both* sets of needs, with equal vigor and dedication.

- A *high-quality* product is one that meets the needs of its customers at the lowest cost (minimum waste and maximum consistency). Quality is measured by determining the extent to which these two criteria are met. Observing the two possible characteristics of the outputs of a process—product features (good characteristics) and deficiencies (bad characteristics)—makes this determination. It follows that the job of managers (leaders) is to produce high-quality products by putting in place all the necessary means to do this. The necessary means are described in the principles of the trilogy.

- *Process outputs* can embody both product features and deficiencies.

- *Product features* are the characteristics, or properties, of a product that meet specific needs of specific customers. For a person who purchases an automobile for personal use, the needs are the benefits sought from the automobile: transportation, style, status, and comfort. The associated product features are what specifically about the product meets each need. For the needs just listed, the product features may be a capacity of six passengers, power, top speed, acceleration, range, aerodynamic design, price, leather bucket seats, and premier sound system. For a patient in a hospital, desired benefits (needs) may include quick admission, comfortable bed, and curing the ailment. The associated (service) features might be pre-admission by phone in advance of checking in, scientifically designed power-adjustable bed, and a recovery with no infection, no relapse, and no repeated treatment. Product features are what make the customer happy and give satisfaction. Product features are what produce sales and repeat demand.

• *Deficiencies* are things that are wrong with the product—or the process. Examples of deficiencies are specific defects, excessive time cycles, and excessive costs. Defects include errors, omissions, the need for reworking or scrapping a product, etc. Defects make customers unhappy and produce dissatisfaction. Defects result in a loss of profitability because of the costs of poor quality (unnecessary waste) that eat into the bottom line and can reduce or prevent sales.

> **NOTE:** Satisfaction and dissatisfaction are not opposites. They are separate dimensions of customer reaction. Reducing dissatisfaction does not produce satisfaction; it merely reduces dissatisfaction. Increasing satisfaction does not decrease dissatisfaction; it merely increases satisfaction. Therefore, you cannot make customers who are dissatisfied by defects in your product less dissatisfied and happier by giving them more product features.

For example, you can't make people happy with their unreliable cell phone service by granting them a large amount of free air time. They'll be pleased with the free time, but they'll still be unhappy with the lousy phone service.

The Juran Trilogy is a management system for producing ideal product features *and* minimum deficiencies at lowest possible costs.

Following the Trilogy permits an organization to *maximize customer satisfaction* (by economically producing ideal product features) and *minimize dissatisfaction* (by reducing or eliminating deficiencies and the costs of poor quality—waste—associated with deficiencies).

Embedded in the Trilogy principles is the concept that work gets done; that is, products are produced by means of *multifunctional* processes. The production of a product is the ultimate result of inputs and outputs at each successive step in a spiral of functional activities. In manufacturing, for example, the spiral may travel through functions such as sales, design, test, procurement, storage and retrieval, production and inventory control, operations, packaging, shipping, customer service, and so on. The production process is theoretically seamless, and travels from beginning to end of the spiral and then repeats the cycle over and over.

Although the spiral is seamless, in practice, it is not always observed to operate that way. When one attempts to observe organizational processes, one encounters some strange sights. One strange sight is that a typical organization is organized into a group of relatively vertical functional silos, each devoted to providing a specific organizational function: procurement, design, sales, production, transportation, training, human resources, and the like. Each function has its own manager, its own budget, and its own goals. It may have its own norms and customs, and its own language. People residing in a given silo look to the manager of that silo for guidance and direction, and have a tendency to view their job in terms of the tasks they perform on behalf of their own function. Often, the organization reward system reinforces all this.

This state of affairs is strange because the processes by which the organization's ultimate products are produced move *across* multiple functions. Nevertheless, it is rare to observe multifunctional management structures to control the whole process as it moves across the fiefdoms of the functional managers. Cost accounting procedures typically track *functional*, not *multi-functional* costs, even though the major costs of poor quality (waste) occur multi- or cross-functionally.

One observes multi-functional processes being managed functionally! All this can result in delays, confusion, bad feelings, unreliable communication, and uncoordinated activity, not to mention unnecessarily high costs and other barriers to efficiency and effectiveness. Fortunately, there is an emerging trend toward multifunctional accounting and managing, but it is by no means universal.

Clearly, management has a job to do to untangle the organization from this disconnect between performance and structure that produces a steady stream of poor quality and unnecessary waste. When managers follow the principles of the trilogy, they can do the job.

THE BASIC TASKS OF MANAGEMENT

According to the Juran Trilogy, management performs three basic tasks.

First, it sets goals (strategic and operational goals and quality goals for products and processes), and puts in place the means to attain those goals. The process for setting organizational goals is called *strategic planning* and *quality planning* (product and process design). The umbrella term "planning" is used to designate activities that are carried out in preparation for taking action. Planning establishes, among other things, specific standards/specifications for specific products or processes. Financial planning, a similar exercise, sets out the financial goals and the means to achieve them and results in a budget (a financial plan).

Second, management prevents or corrects unwanted "bad" change. This process is known as *control*. More precisely, control consists of measuring actual performance, comparing it to the target or standard, and taking action on the (bad) difference. Control maintains the standards, requirements, and specifications set in planning. The goals of control are stability, repeatability, and consistency.

Third, management creates unprecedented "good" change. This process is called *breakthrough*. Breakthrough is a deliberate change, a dynamic, decisive movement to unprecedented higher levels of organizational performance than is currently in the plan and maintained by current controls. Breakthrough results in improved standards and specifications.

Planning, control, and breakthrough are each essential for organizational vitality, but individually they are insufficient. Therefore, *managers must accomplish all three*. Each arises from a specific prerequisite for organizational survival. Each fulfills a specific vital purpose and function. Each entails its own distinctive sequence of events, tools, and techniques. Each of the three interacts with the other two. Without planning, control, and breakthrough, an organization would be incapable of prolonging its existence. Put plainly, no organization can survive if these functions are not performed.

In carrying out their managerial tasks of planning, control, and breakthrough, managers basically make economic decisions on behalf of the welfare of the organization. Accordingly, we will discuss the three basic managerial tasks in terms of costs, specifically the costs of unplanned waste, the so-called costs of poorly performing processes (COP^3). The higher the COP^3, the lower the bottom line. These costs pay for the non-value-added activities an organization is forced to perform when compensating for any type of deficiency. The discussion moves through the trilogy, starting with planning, then discussing control, then breakthrough.

COSTS OF POORLY PERFORMING PROCESSES (COP^3) AND THE TRILOGY

The cumulative costs associated with poor performing products and processes can easily consume your organization's profits. A traditional common metric of poor organizational performance is the "costs of poor quality" (COPQ). COPQ are the costs of unplanned, unnecessary waste. COPQ has been used as an ad hoc financial report to quantify the losses due to poor performance. Historically, studies have shown costs of poor quality to run as high as 15 to 40 percent of costs of goods sold or about 15 to 20 percent of sales revenue, an extraordinary sum of money lost. These accumulated costs are usually not reported to management, at least not in an easily understood financial manner that would permit making decisions for countermeasures. Managers remain unaware of their full extent and impact because traditionally, calculations of costs are made by comparing overall departmental expenditures with overall departmental budgets.

In this book, we want to provide an updated way to measure and think about traditional COPQ. We call this new measure "COP cubed." COP^3 means "Costs of Poorly Performing Processes." It seems to be an easy way to convey the real meaning of COPQ and the purpose for calculating COPQ.

Information about overall costs, as conveyed by traditional accounting reports, does not reveal enough detail to enable managers to identify specific problems and opportunities for improvement as they relate to the way work gets done— through cross- or multi-functional processes. Furthermore, in addition to departmental activities that are carried out because of poor performance, there are numerous activities (such as repair, rework, clarifying unclear instructions, etc.) that are performed to compensate for poor quality, as processes move across and between the various departments. None of these activities would be carried out if there were no quality problems. For example, we wonder whether most of the staff in a traditional Accounts Receivable Department would still have jobs if the reasons clients don't pay were to be discovered and eliminated!

> **NOTE:** A *quality problem* is a specific failure to meet a specific need of a specific customer. There is no such thing as quality in general, except in the sense of the aggregate extent to which each vital need of each vital customer is met. Hence, one cannot improve quality in general. One improves upon specific failures to meet specific needs of specific customers.

With a COP³ analysis as a starting point, important problems are easily identified, selected for projects, and eliminated.

Intra- or cross-functional activities, largely hidden from detection by traditional cost accounting systems, can be extremely expensive, adding significantly to the costs of poor quality, and reducing significantly the bottom line. A recent report provided by the Midwest Business Group on Health and the Juran Institute showed that the total costs of poor quality in the United States health care system are approaching 30 percent of the $1.3 trillion spent annually on health care.

Organizations rarely have departmental budgets that show costs of activities in enough detail to pinpoint specific improvement opportunities. Even less common are cross- or multi-departmental budgets or measures for inter- or cross-departmental costs, such as those provided by activity-based accounting systems. Consequently, management is generally unaware of the extent of the opportunities—indeed of the necessities—for making specific improvements in performance, and therefore, they do not take place. Serious quality problems, with their associated steady but avoidable drain on resources and reputation, remain unaddressed.

COPQ would all disappear if every task was always performed without deficiency. COP³ are the costs of unplanned, theoretically unnecessary waste. Breakthrough improvement can capture these costs from beginning to end of the supply chain. This would drive costs down by targeting and removing the deficiencies that caused the excess costs in the first place.

There are three major categories of COP³:

- Appraisal costs
- Internal process failure costs
- External product failure costs

APPRAISAL COSTS

The portion of appraisal costs that are included in COP³ are those costs associated with discovering deficiencies before external customers are affected by them. Some of these costs may be excessive and therefore "non-value added." Appraisal costs include costs of checking and inspecting your processes and work in progress to assure that the processes are performing to standard, and the products meet the requirements.

Examples of Appraisal Costs

- Testing appliances before shipping
- Reviewing insurance policy before mailing
- Inspecting purchased equipment/supplies
- Proofreading reports or correspondence
- Auditing customer bills prior to sending bill
- Testing an automobile to be certain repairs were made

Note that these activities are performed because we expect to find something bad. We don't trust our suppliers or our processes, probably with good reason. The portion of inspection costs incurred to carry out routine quality control audits are not costs of poor quality. This is because these audits are supposed to occur anyway. They are part of one of the three basic managerial tasks: control. (Control is described in detail in Chapter 4, "The Control Processes.")

INTERNAL PROCESS DEFICIENCY COSTS

Internal process deficiency costs are the costs to repair, replace, or discard defective work that the customer does not see directly, although customer service may well be adversely affected, if the deficiencies delay delivery time. Internal deficiencies usually accumulate as a process moves from task to task, step to step. In Six Sigma, the rolled throughput yield metric is a good example of a measure of deficiency cost buildup, as shown in Figure 2.1.

Examples of Internal Process Deficiency Costs

- Working overtime to make up for schedule slippage
- Taking another X-ray
- Replacing metal stampings that do not meet specifications

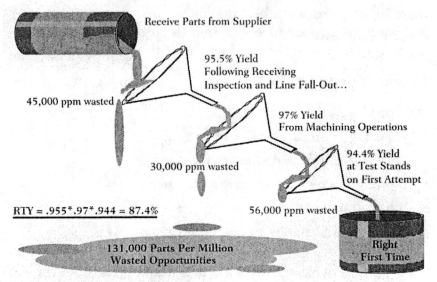

Receive Parts from Supplier

95.5% Yield
Following Receiving
Inspection and Line Fall-Out...

45,000 ppm wasted

97% Yield
From Machining Operations

30,000 ppm wasted

94.4% Yield
at Test Stands
on First Attempt

RTY = .955*.97*.944 = 87.4%   56,000 ppm wasted

131,000 Parts Per Million
Wasted Opportunities

Right
First Time

FIGURE 2.1   Rolled throughput or total process yield.

- Repainting scratched surface of products
- Making up for unplanned downtime
- Replacing products damaged during moving, packing, or shipping
- Rewriting part of a proposal
- Correcting errors in various databases
- Stocking extra parts or components to replace predictably defective ones
- Scrapping products that do not meet specifications

EXTERNAL PRODUCT DEFICIENCY COSTS

External failures occur after the product leaves our organization or when a service is performed. These are the failures that customers, regulators, and society see and feel as a result of coming into contact with your product. They are the most expensive to correct, and they are costly in other ways as well. They result in extra costs of attempting to regain the customer's confidence, and they can easily result in a loss of customers—a cost that usually cannot be calculated.

Examples of External Product Deficiency Costs

- Satisfying warranty claims
- Investigating and resolving complaints
- Giving credits and allowances to customers
- Offsetting customer dissatisfaction with a recovery strategy
- Collecting bad debts
- Correcting billing errors
- Expediting late shipments
- Replacing or repairing goods damaged or lost by carrier
- Housing stranded passengers from a canceled flight
- Paying interest or losing discount for late payments
- Giving onsite assistance to customers to overcome field problems
- Going into litigation, due to alleged mistreatment of a patient

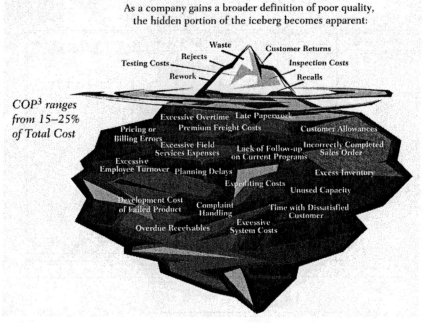

FIGURE 2.2   Cost of poorly performing processes (COP³).

Note how traditional cost accounting and reporting practices "hide" many costs of poorly performing processes. Just as most of the ice in an iceberg is invisible from the surface of the water, a very large proportion of costs of poorly performing processes are absent from traditional financial reports. This invisibility may explain why so many organizations continue to tolerate such high levels of avoidable costs. In effect, they are *not* really tolerating them; they are simply ignorant of them.

With this background, we continue our description of the three basic managerial tasks by referring to Figure 2.3, the Juran Trilogy.

The vertical (Y) axis represents COP^3 as a percent of revenue. Anything that goes up is bad. The horizontal (X) axis represents time marching on from time zero, at the left, when operations begin.

PLANNING

Starting at the left side of Figure 2.3, before time zero, strategic planning and deployment have already occurred. Quality

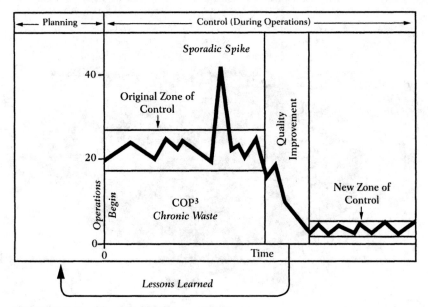

FIGURE 2.3   The Juran Trilogy.

planning (product and process design) has also already been completed in preparation for operations. Quality planning establishes the design of a product that will meet customers' needs, together with the process that will produce the product.

Quality planning follows a more or less universal sequence of steps, as follows:

- Identify customers and target markets.
- Discover customer needs.
- Translate needs into our language: operational definitions, standards, specifications, etc.
- Develop a product (could be a service) that meets customers' needs.
- Develop a process that will create the product in the most economical way.
- Transfer the plan to the operating forces.

Major customers and their vital needs have been identified, products have been designed with features that meet the customers' needs, and processes have been created and put into operation to produce the products with the desired product features. Practitioners of Six Sigma will recognize this process as a fundamental way of describing the define-measure-analyze-design-verify sequence employed in Design for Six Sigma (DFSS). (A discussion of Quality Planning, including DFSS, is described in Chapter 3, "The Planning Processes.")

As operations begin, it becomes apparent that some features of the product or process design are in error, have been overlooked, or have been poorly executed. The yield of operations in this particular example fluctuates around 22 percent COP[3]. This is the chronic level. What is the source of this chronic level? In a sense, it was planned that way, not on purpose of course, but by error or omission.

No matter what they try, the flaws in product or process design make it impossible for the operating forces to do any better, on average. (This situation is rendered even more intractable because in traditional organizations, the operating forces have no skill in, or responsibility for, product or process design or redesign.) As time passes, the chronic level of waste becomes regarded as the norm. ("That's how it is in this business.") Some think of it as a fate that is not susceptible to improvement. With the passage of more time, almost by default, the "norm" eventually becomes designated as the "standard," and the associated costs of poor quality are unknowingly built into the budget! In our example, unless the COP^3 does not in effect exceed 22 percent, performance is not regarded as abnormal or bad, or exceeding the budget! The organization has thus desensitized itself from recognizing its major opportunities for bottom-line-boosting improvements. (Dr. Juran would say that the management alarm system has been disconnected.) Furthermore, the cost accounting system usually does not provide managers with complete information about the levels of COP^3. So it is not surprising that defects are not addressed and corrected.

CONTROL

To return to Figure 2.3, when the COP^3 sporadically flares up to unusually high levels in some locations, this triggers troubleshooting or quality control. Control consists of measuring actual performance, comparing actual to the standard, and taking action on the (bad) difference.

Control employs the feedback loop, which results in corrective action. The elements of the feedback loop (and hence of control) are (1) the *control subject*, which is the process or product characteristic to be controlled; (2) a *sensor*, which measures the control subject; and (3) an *umpire* who receives the measure from the sensor, and then consults (4) the *standard* (expressed in units of measure). If the standard is not met, the umpire energizes (5) an *actuator*, who adjusts the process to bring it back into compliance with the standard. The actuator may be a device, a supervisor, or persons doing the work being controlled.

Example of Control

The thermostat.   The thermostat is an example of an automated feedback loop, where all the elements of the feedback

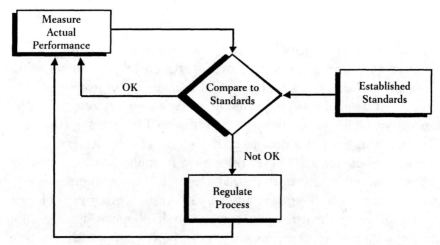

FIGURE 2.4 Control feedback loop.

loop, except establishing the standard, are performed automatically by means of a device or devices. (1) The *control subject* is ambient air temperature. (2) The *sensor* is a thermometer that measures temperature. (3) The *standard* is whatever temperature at which the thermostat is set. (4) The *umpire* consists of mechanical contacts. As temperature increases beyond the temperature setting, contact is made, which energizes the (5) *actuator* (a circuit breaker) and the heat is shut off. When temperature falls below the temperature setting, the contact causes the heat to be turned back on.

Control is a form of improvement, in the sense that a bad situation—failing to meet the standard—is made better by re-attaining the current standard, but it is not breakthrough improvement. It is merely removing unwanted change and restoring the process to meet the current standard. (A discussion of control is contained in Chapter 4, "The Control Processes.") Control is carried out by a more or less standard sequence of events:

- Evaluate actual performance.
- Compare actual performance with the standards.
- Act on the (bad) difference.

Attaining a better new standard of performance requires breakthrough in the chronic level of performance.

BREAKTHROUGH

In contrast to control, which seeks a stable state of conformance to current standards, breakthrough improvement seeks to create unprecedented beneficial change by improving on the standard itself. In our experience, once managers recognize the extent of COP^3 (usually by performing a COP^3 study), their dissatisfaction with the results of the study usually leads them to make a conscious deliberate decision to make lasting improvements by setting in motion breakthrough improvement.

"Breakthrough improvement" is one component of the Juran Trilogy. It refers to meticulously diagnosing the root causes of chronic, mysterious, costly problems with existing products or processes; devising remedial changes that remove or go around the causes; and implementing new controls so the problem cannot return.

> NOTE: "Going around the causes" refers to the situation when no change to the process or no technology is capable of removing a cause. In this case, it becomes necessary to make the symptoms of the problem "robust" to the cause, that is, unaffected by the cause. Accomplishing this can be challenging, and sometimes may not be possible. Nevertheless, the attempt is made.

On Figure 2.3, breakthrough improvement is represented by the dipping curve at the right. The diagram depicts a one-time breakthrough improvement effort that results in a gain in performance that goes on and on, perpetuated by the installation of new controls that prevent performance from returning to its previous not-so-good level. At the start of a breakthrough improvement effort, the desired remedy is not known. It is not even knowable, because the cause(s) of the problem are unknown. The problems that become candidates for breakthrough improvement are the chronic ones. (Sporadic problems are addressed by control.) They are chronic because they are mysterious. No one really knows the cause(s). Many will say they do, but rarely can breakthrough improvement project teams predict either what cause(s) the team will discover, or what remedial improvements will ultimately be designed and put into place. The work of a breakthrough improvement project team can be likened to the work of a detective squad: solve a mystery by assembling evidence that proves who are the guilty parties (what are the causes) and bring this evidence to court (management) for appropriate action.

Breakthrough improvement is carried out by a universal sequence of events, as follows:

Breakthrough Improvement

- Identify a problem (something wrong with a product, service, or process).
- Establish a project.
- Analyze symptoms to establish precise knowledge of baseline performance.
- Generate and test theories as to what are the causes of the analyzed symptoms.
- Identify root causes.
- Develop remedial improvements—changes to the process that remove or go around the causes.
- Establish new controls to prevent recurrence and to hold the new standards.
- Deal with resistance to change.
- Replicate the result.

Practitioners of Six Sigma will recognize this as a fundamental way of expressing the Define-Measure-Analyze-Improve-Control sequence followed in Six Sigma DMAIC Improvement. (Six Sigma and other approaches to breakthrough improvement are described in Chapter 8, "Breakthroughs in Current Performance.")

Note that in following this sequence, much attention is paid to understanding thoroughly and in detail how the current "bad" process actually performs. Then, and only then, is an attempt made to understand what causes the process to produce the defects under investigation. When, and only when, evidence has been accumulated to prove what are the root causes, remedial changes to the process that produced the problem are designed to remove or go around the causes. No remedies or improvements are even discussed, much less designed and implemented, until the root cause(s) of the problem are scientifically established beyond doubt.

Effective quality improvement discovers *root* causes before applying a remedy. Like peeling an onion, layer after layer of causation is revealed until the root cause is visible. When action is taken to remove or go around a root cause, many improvements happen, because many problems or symptoms of a problem have common cause(s). Extracting root causes instead of more superficial causes pays off many times over by reducing the time and effort the discovery journey takes and eliminating the need for repeated iterations of searching.

EXAMPLE OF DISCOVERING ROOT CAUSES

Here's an example of what we mean. The stonework in the Jefferson Memorial located in Washington, D.C. was crumbling in places, especially inside the structure. A quality improvement project team decided to go after the root cause of this.

> **NOTE:** If we tell you at this point that the ultimate remedy that stopped the stone from crumbling was to turn on the outside illuminating lights *after* sunset, instead of at dusk, would that sound crazy? It probably would have to the project team, too, at the beginning of the project, before it went beyond considering superficial causes, and penetrated successive layers of causation down to the root cause layer.

Let's follow the search for root cause, layer by layer. The question was: "What is causing the stone to crumble?" Well, it was the frequent washing with strong detergent that gradually dissolved the stone.

So the next question was: "Why are there so many washings?" Well, it was all the bird droppings that rapidly accumulated.

The next question was: "Why are there so many bird droppings?" Well, it's because of all the birds that come into the structure to eat the enormous number of succulent spiders who have webs all over the place.

"Why are there so many spiders?" Well, it's because of the almost endless supply of midges that the spiders eat.

"Why are there so many midges? Where do they come from, and what are they doing here?" Well, at dusk each night, the midges hatch and emerge from the nearby water basin. The bright lights that illuminate the monument attract them.

Root cause: These lights are turned on at dusk, just as the midges emerge. When they approach the lights, they are caught by the awaiting spiders.

So the root cause was the time when the lights were turned on.

Remedy: Turn the lights on a while after dusk. By then the midges are dead or dying. The food chain is broken. The spiders and the birds depart. No more washings. Problem solved.

Note that the problem would not have been solved by removing the most superficial cause—the strong detergent. A weaker detergent would probably lead to even more washings, and no fewer midges.

The sequence goes further, to assure that breakthrough is sustained. New controls, particularly on the remedial changes made to the process being improved, are designed and implemented. Periodic audits are performed to ascertain that the controls are actually being applied.

INTERACTIONS OF ELEMENTS OF THE TRILOGY

Recall that the output of any process has two aspects:

1. Product features
2. Product or process deficiencies

Product features come from quality planning. Quality planning occurs when what is needed currently does not exist. It must be created (such as designing a new product).

Freedom from deficiencies comes from breakthrough improvement. Breakthrough improvement occurs when an existing product or process fails to meet specific needs of specific customers. It must be fixed. Not enough is known about the cause(s) of the deficiencies to simply fix them. The causes must first be discovered before remedies can be applied.

Control is an integral part of both quality planning and breakthrough improvement. Controls are designed toward the end of each process to assure the gains are sustained long after the design or remedy is implemented. In quality planning, controls are designed and put into operation on all the vital transactions and operations required to meet the standards built into the product and process design. In breakthrough improvement, controls are designed and put into operation on the remedial changes to the process that constitute the improvement.

Controls are the means to assure that a given design or breakthrough project has to be executed only once to get results. When controls are properly and consistently applied, the product and process will work as designed and will meet standards, and the original problem solved by breakthrough cannot return. If controls are missing from a breakthrough project, or inconsistently applied to a reversible remedy (for example, a change in procedure where an operator can regress back to the comfortable old ways by not following the new procedure), the benefits of the remedial gains can be lost. The entire breakthrough project becomes a wasted effort.

Historically (in the 1970s and before), a breakthrough meant, say, a 10 percent improvement in the performance of a process to bring the process down to acceptable quality levels (AQLs; measured in defects per hundred—percentages). Defect levels of 5 to 10 percent were commonly regarded as acceptable. In today's competitive environment, breakthrough may require achieving improvement in orders of magnitude, bringing deficiency levels down to a few defects per million! (Six Sigma = 3.4 defects per million opportunities for making a defect.)

HIGH POINTS OF "THE JURAN TRILOGY"

- The purpose of an organization is to meet the needs of its customers at the lowest optimum cost.

- Products can be goods, services, or information.
- Products are produced by processes.
- A process is a sequence of events or tasks that creates an output: a product. A process includes everything involved at each step: people, techniques, equipment, materials, facilities, etc.
- A customer is one who receives the output of a process or any step in a process.
- A client is a special kind of customer who pays for what is received.
- An external customer is one located outside your organization.
- An internal customer is one who is located within your organization.
- Meeting the needs of internal customers is a prerequisite for meeting the needs of external customers.
- Management must meet both sets of needs.
- A high-quality product is one that meets the needs of customers at the lowest optimum cost.
- Quality is measured by the extent to which specific needs of specific customers are met.
- Process outputs can embody both product features and deficiencies.
- Product features are the characteristics of a product or service that meet specific needs of specific customers. Product features produce satisfaction.
- Deficiencies are things that are wrong with the product or the process. Deficiencies produce dissatisfaction.
- Satisfaction and dissatisfaction are not opposites. They are separate dimensions of customer reaction to a product.
- Management fulfills three basic functions: planning, control, and improvement.
- Planning consists of setting quality goals and putting in place the means to reach those goals. Common activities: product and process design.

- Control consists of preventing and/or correcting unwanted, unplanned bad change in a process.
- Improvement consists of the purposeful creation of beneficial change.
- All three are vital to the survival of an organization. Together, they form the "Juran Trilogy."
- Following the Trilogy permits an organization to maximize customer satisfaction and minimize dissatisfaction.
- Costs of poorly performing processes (COP^3) are the costs of avoidable unplanned waste in a process.
- CPO^3 would be reduced if all elements of the trilogy were always carried out properly.
- The resulting savings would go directly back into the budget (not-for-profit) or to the bottom line (for-profit).

THE PLANNING PROCESSES

TACKLING THE FIRST OF THE TRILOGY ELEMENTS: PLANNING

Your ability to satisfy your clients and customers depends on the planning process because the goods you sell and the services you offer originate there.

The planning process is the first of the three elements of the Juran Trilogy. It is one of the three basic functions by which management assures the survival of the organization in its care. The planning process creates designs of products (goods, services, or information) together with the processes—including controls—to produce the products. When planning is complete, the other two elements—control and improvement—kick in.

We will present two versions of the planning model: first, the classic "Juran" version, and second, the more recent statistically and computer-enhanced version known as Design for Six Sigma (DFSS).

The "classic" model is especially useful for designing or redesigning processes or relatively simple products economically. The authors have witnessed the design of superb products, processes, and services using this model.

Examples include: a prize-winning safety program for a multiple-plant manufacturer; an information system that enables both sales and manufacturing to track the procession of an order throughout the entire order fulfillment process so customers can be informed—on a daily basis—of the exact status of their order; and a redesigned accounts receivable system much faster and more efficient than its predecessor.

The DFSS model is the classic model enhanced by the addition of laptop computers and statistical software packages, which permit the utilization of numerous planning tools not easily used without a computer. The Six Sigma model is suitable for designing even complex products and for achieving extraordinary levels of quality. Although it is time-consuming and expensive in the short term, when executed properly, it produces a healthy return on investment.

THE CLASSIC "JURAN" MODEL OF QUALITY PLANNING

WHAT IS QUALITY PLANNING?

Modern, structured quality planning is the methodology used to plan both product features that respond to customers' needs, and to plan the process to be used to make those features. Quality planning refers to the product or service development processes in organizations. Note the dual responsibility of those who plan: provide the *features* to meet customer needs, and provide the *process* to meet operational needs. In times past, the idea that product planning stopped at understanding the features a product should have was the blissful domain of marketers, salespeople, and research and development people. But this new dual responsibility requires that the excitement generated by understanding the features and customer needs has to be tempered in the fire of operational understanding. That is, can the processes make the required features without generating wastes? Answering this question requires understanding both the current processes' capabilities and customer specifications. If the current processes cannot meet the requirement, modern planning *must* include finding alternative processes that are capable.

Begin at the beginning. The trilogy points out that the word quality incorporates two meanings: first, that the presence of product features creates customer satisfaction, and second, that freedom from deficiencies about those features is also needed. In short, deficiencies in product features create dissatisfactions.

- Removing deficiencies is the purpose of quality improvement.
- Creating features is the purpose of quality planning.

Kano, Juran, and others have long ago agreed that the absence of deficiencies, that is, no customer dissatisfaction, *may not* lead us to the belief that satisfaction is thus in hand. All we can readily conclude is that dissatisfaction goes down as deficiencies are removed. We cannot conclude that satisfaction is therefore going up, because the removal of irritants does not lead to satisfaction, it leads to less dissatisfaction. It is only the presence of features that creates satisfaction. Satisfaction and dissatisfaction are not co-opposite terms.

It is amazing how many companies fail to grasp this point. Let's take for example the typical "bingo card" seen in many hotels. These are replete with "closed-ended" questions. They ask, "How well do you like this on a scale of 1–5?" for example. They are not asking, "How well do you like this?" This is the exact co-opposite of, "How well don't you like it?" Therefore, any so-called satisfaction rating that does not allow for open-ended questioning such as, "What should we do that we are not already doing?" "Is there someone who provides a service we do not offer?" will always fall into a one-sided dimension of quality understanding. What then does a composite score of 3.5 for one branch in a chain of hotels really mean compared to another branch scoring 4.0? It means little. Their so-called satisfaction indices are really dissatisfaction indices.

So we arrive at the basic fundamental of what quality really is. In the fourth edition of *Juran's Quality Handbook*, the authors adopted a definition that Juran had postulated long before: quality means fitness for use. Let's explore this concept.

First, the definition of "fitness for use" takes into account both dimensions of quality—the presence of features and the absence of deficiencies. The sticky points are these: Who gets to decide what "fitness" means? Who decides what "use" means?

The user decides what "use" means, and the user decides what "fitness" means. Any other answer is bound to lead to argument and misunderstanding. Providers rarely win here. Users, especially society at large, generally always win.

Take yourself as a consumer, for example. Did you ever use a screwdriver as a pry bar to open a paint can? Of course you did. Did you ever use it to punch holes into a jar lid so your child can watch bugs? Of course you did. Did you ever use it as a chisel to remove some wood, or metal, which was in the way of a job you were doing around the house? Of course you did. Now wait just a moment…a screwdriver's intended use is to drive screws!

So the word "use" has two components, intended and actual. When the user uses in the intended way, both the provider and the user are happy. Conformance to specification and fitness for use match. But what about when the user uses in the nonintended way, like in our screwdriver example? What then, regarding specifications and fitness?

Look at it like this:

	FIT FOR USE? YES	No
Conform to specifications? Yes	No real problem for either provider or user.	Provider does not understand "fit."
No	Provider does not understand "use."	No real problem for either provider or user.

To delve even deeper, how does the user actually use? Here we find another juncture: the user can create artful new uses.

"2,000" Uses for WD-40

WD-40 was formulated years ago to meet the needs of the U.S. space program. Not many know the origins of the brand name. "WD" refers to water displacement. "40" is simply the 40th recipe the company came up with. But as the product moved into the consumer market, all kinds of new uses were uncovered by the users. People claimed it was excellent for removing scuff marks from flooring. They claimed it could easily remove price stickers from lamps, inspection stickers from windshields, and bubble gum from children's hair. The company delighted in all of this.

But the company didn't release all of those clever new uses for public consumption. People also claimed that if they sprayed bait or lures with it, they caught more fish. Those with arthritis swore that a quick spray on a stiff elbow gave them relief. Let's not go too far....

What about use where the product obviously cannot work? The Latins had a word for this: AB-use (abuse), where the prefix "ab" simply means "not."

NONINTENDED	INTENDED
Clever, unplanned for, discovered by user	Provider and user in sync
Abuse	

Some examples will help; back to the screwdriver. You could argue that using the screwdriver as a pry bar, chisel, or punch is abuse. But, clearly, many U.S. manufacturers have provided a product that can withstand this abuse, and so use then falls back into the "intended" column (whether this came as a result of lawsuits or from some other source). Further, a look at commercial aircraft "black boxes," which are international orange, by the way, shows that they clearly survive in circumstances where the aircraft do not survive.

Understanding of use in all its forms is what modern planning seeks to achieve.

Last, modern planning, as we shall see over and over, seeks to create features in response to understanding customer needs. We are referring to *customer-driven features*.

A different type of product planning, where features meeting no stated need are put out for users to explore, is beyond the scope of this chapter. Examples such as 3-M's Post-It Notes and the Internet are items where we collectively did not voice needs, but which we cannot imagine life without, once we embraced their features.

RELATIONSHIP TO THE JURAN TRILOGY

Clearly, the Juran Trilogy, explained in the previous chapter, has a direct bearing on quality planning. Indeed, in his first description of the trilogy, Juran showed the linkage between planning, control, and improvement. For the purposes of this chapter, we can reflect on planning in two ways: strategic and tactical.

Strategic planning (some call it strategic quality planning) is explained in Chapter 11, "Strategic Quality Planning and Deployment." It is a process for establishing long-term goals. For this chapter on quality planning (a process for creating new products and processes), we need to stress that product planning must be congruent with organizational strategy. Stated another way: product planning is the tactical deployment of strategic intent.

A look at the development of the first Ford Taurus makes the point succinctly. To paraphrase the Ford Taurus story, the company set out to make a four-door family sedan, which would sell for $16,000 [in then-dollars], be the Best-in-Class, and have a "platform" that would see the company through 10 years of manufacturing and design changes. Of course, many car companies, including Ford, continually probe the market with concept cars and concept features. "Making it to market" is problematic at best. Many concepts die on the fabulous auto show floors. Those that do launch without clearly meeting an understood need almost always have poor market results. In the case of the Taurus, and many other marques that are developed by disciplined quality planning (customer-driven features), we see sustained market share.

WHAT IS A PROCESS? WHAT IS A PRODUCT?

Up until this point, we have not been clear as to the meaning of the words "process" and "product." Let's examine the following situation familiar to many:

A Look at the Checkbook Balancing Process

What's a process anyway? If we begin with the notion that all work is a process (for example, driving to the store, writing a report, making a machined part, preparing an invoice...) then we must settle on what we mean by "process."

Juran said it best: A process is "a systematic series of actions directed to the achievement of a goal." The goal, the output, is the product.

So for our little examples, the goals/products might be getting to the store safely, the completed report, the finished part, or the completed invoice. Looking a little deeper, we will use the checkbook balancing process, a chore familiar to many.

First, all processes start with inputs, and all processes create outputs (products). Long ago, the "total quality management" world agreed that the inputs come from "suppliers," and the outputs go to "customers." No more debates about those things anymore.

Then, we found that there are five things cardinal to any process. They are found in each and every process you can bring to mind. These are that all processes:

1. Use machines or some sort of tools or equipment,
2. Employ humans in some degree,
3. Convert or use raw materials in some way,
4. Have a procession of events, usually written as procedures, "SOPs," flow charts, or the like, and
5. Consume or convert energy as the procession of events takes place.

With this in mind, let's look at the checkbook balancing process.

- What are the inputs? The bank statement, along with canceled checks, possibly.
- What are the outputs/goals/products? A balanced checkbook, the output, the goal, and the product.
- Any machines or tools? Calculators *and* pencils.
- Raw materials? Paper and erasers.

- Energy? Batteries for the calculator, heat for the coffee, and brain energy.

- Procedure? There is no bank in the United States that doesn't provide the process steps on the back of each and every page of the statement.

- People? You. You are both the operator and the customer. As both, you want to meet the goal, a balanced checkbook. (The bank or your payees really don't care if your checkbook ever balances.)

Finally, please note that the people in a process generally know the process goals; they don't have to be exhorted. But the other elements of any process are not human, and they go about their way doing whatever it is they do. In our checkbook balancing process, those other elements seem to border on being evil. Many people resort to all sorts of reconciliation tricks; they insert "to adjust" and some numbers to make the whole thing come to a merciful end, for example.

THE SKELETAL OUTLINE OF STRUCTURED QUALITY PLANNING

We have only skirted the nature of "structured quality planning" up until this point. It is time to lay out the universal sequence of activities necessary to plan a product properly (Figure 3.1). They are:

The Universal Product Planning Sequence

1. Establish a project and set design goals.
2. Identify the customers.
3. Discover the customers' needs ("voice of the customer" [VOC]).
4. Develop the product and process features.
5. Develop process controls and transfer to operations.

We shall look at each of these as we step through the sequence at a high level.

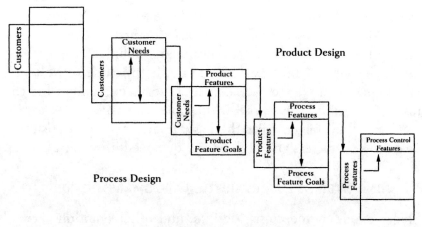

FIGURE 3.1   The universal sequence of activities necessary to plan a product.

PLANNING STEP 1: ESTABLISH A PROJECT

As stated in a previous chapter, and paraphrased here in the planning context: All planning takes place project by project.

There is no such thing as planning in general; there is only planning in specific. In strategic planning, we set out the vision, mission, strategies, objectives, and so on. Each is a specific thing. In product planning, we start with a project, that is, something to plan. We might be planning a new training room, a new car, a wedding, a customer toll-free hotline, or a new Internet process for bidding on travel booking (such as Priceline.com). Note that each is a specific thing, and each can be clearly differentiated from anything else. A training room is not a cafeteria, a new car is not a Howitzer, a hotline is not long distance service, and the travel booking process is not a book-store online. This is a significant point. Without being able to differentiate what we are planning from anything else, every-thing collapses into vagueness. So a project is our starting point.

PLANNING STEP 2: IDENTIFY THE CUSTOMERS

Going back to the 1980s total quality management (TQM) days, we learned that those who receive the product are cus-tomers in some way. If we were designing a training room, the trainees would be an important customer segment. So too would the custodians, because they have to clean the room, set it up different ways, and so on. Customers of the new car include the purchasers, the insurance companies, the dealers, the carriers, etc. Customers of the hotline include our clients, our service agents, etc. We can include as customers for the travel process the travelers, airlines, and the Web server entity.

From all of this emerges the basic understanding: A customer is a cast of characters and each has unique needs that must be met.

PLANNING STEP 3: DISCOVER THE CUSTOMERS' NEEDS

Wants, needs, perceptions, desires, and other emotions are all involved in our discovery of customer needs. We need to learn how to separate things and prioritize them. But at this point, we need to emphasize that not all high-priority customers (like the car buyer) are the only ones with high-priority needs. We also stress that just because some customer entity is lower in priority doesn't mean at all that it automatically has lesser-priority needs. We need to understand the "voice of the customer" and the "voice of the market."

Take for example the automobile carriers; we simply cannot overlook their needs for the car to be only so high, and only so wide. If we ignored their needs, they could stop the product from reaching the cash-paying ultimate customer, our buyer. So too could regulators (the various states, the National Highway Transportation Safety Board, the Environmental Protection Agency, etc., impose "needs" which if unmet, could stop the process from going forward at all).

So from all of this, we reach another point: Customers have to be prioritized in an agreed-upon way.

PLANNING STEP 4: DEVELOP THE PRODUCT FEATURES

The word "feature," as used in product planning, means what the product does, its characteristics, or its functionality. In structured product planning, we adopt a different definition: A feature is the thing that the customer employs to get his/her needs met.

For example, in our training room, the trainees need to take notes as they learn. A feature might then be a flip chart, a white board, or a desk. Our custodians might need to move things around quite a bit, so features might include portability, size, weight, and modularity.

As our list of features grows, we soon realize that we cannot possibly have all features at the same priority level. So we need a way to put things in order, once again, and in an agreed-upon way. We finalize by optimizing and agreeing on the list of features, and the goals for them as well. Note that optimization means: Not all features survive product planning.

PLANNING STEP 5: DEVELOP THE PROCESS FEATURES

Because we know that the process is the thing that creates the features, we need to examine current and alternative processes to see which ones will be used to create the product features. We need to be sure that the product feature goals can be accomplished via the processes we choose. In other words: Process capability must reconcile with product requirements.

That statement is very important. No process knows its product goals; product goals come from humans. Ideal product goals would naturally reflect the various customers. But the key issue is: Variation comes from processes; goals come from humans.

In the example of the training room, process goals might be to re-set the room in 20 minutes, keep a supply of flip charts in a closet, certify the trainees to a standard, and so on.

As before, we need to list all the possible routes to making the product, select the ones we will use based on some rationale, establish goals for the processes, and reach an optimum.

PLANNING STEP 6: DEVELOP PROCESS
CONTROLS/TRANSFER TO OPERATIONS

Develop process controls.   Control is basic to all human activity, from how the body regulates itself as to temperature and metabolism, to financial controls in how we run our organizations or homes. Control consists of three fundamentals:

· Knowledge of what to do

- Knowledge of how to do it
- Knowledge of the result, and what to change to make the result happen consistently

This set of fundamentals is enduring in concept, from long-passed civilizations' stonecutters and artisans, through the development of the craft era, to today's concept of self-control (see the chapter on Control).

In product planning, we need to ensure that the processes work as designed, within their capabilities. In the training room, controls might take the form of a checklist for re-setting the room, and a minimum inventory of flip charts, for example.

Control makes use of the concept of the feedback loop. Here's an example you might keep in mind:

Did you ever check the oil in your car? The dipstick is a form of control chart. Note that we begin with a control subject (volume of oil), a unit of measure (quarts/liters), a sensor (you and the dipstick), and a goal (keep the oil somewhere between "full" and "add"—inside those hash marks). We then move on to sample the process (clean the dipstick, put it back in, remove it, and observe the oil level). We then adjust when adjustment is called for (oil levels below "add" require us to add oil until we bring the oil up to somewhere between add and full, the agreed goal). If the oil is already in the hash marks, the control activity is to replace the dipstick, shut the hood, and drive on until another checkpoint is reached (perhaps next month).

Note that the control activity must reflect the agreed-upon goal for control. In the engine oil example, the control point was "inside the hash marks," so the control action is to bring the oil to somewhere "inside the hash marks." Many people miss this point; they add oil until the stick reads "full," for example. This is overcontrol. Control actions must reflect control goals

Transfer to operations.   Transfer to operations winds up the whole planning process. As used here, "operations" means those who run the process, not "manufacturing." To continue the examples used earlier, "operations" for the training room is the activity of the trainers, the custodians, and the purchasing department. For the new car, "operations" includes manufacturing, transport, dealer relations, and the legal department. For the hotline, "operations" means the customer service agents who answer the phone. In the travel bidding process, "operations" includes those who shop the bid or reject it, and those who maintain the software that interfaces the prospect with the carriers.

From the lessons of the era of productivity, the Industrial Revolution, and into the twentieth century, we have learned that the involvement of the "operators" is key to any well-running process.

With the development of the Ford Taurus came solid understanding of the value of a "platform" team. Designers, engineers, workers, purchasing agents, salespeople, and managers all sat under one roof to develop the car. The concept of platform teams is well-ingrained in many car companies today. The Chrysler Technical Center in Auburn Hills, Michigan, is a later example of such broad collaboration. Thus, successful transfer to operations must include the operators in the planning process as early as possible.

THE RELATION OF THE QUALITY PLANNING STEPS TO THE FREQUENTLY USED TOOLS

- *Affinity diagrams.* A diagram that clusters together items of similar type; a prelude to a cause–effect diagram used in quality improvement; and used in quality planning to group together similar needs or features.

- *Benchmarking.* The technique of openly sharing and investigating the best practices of organizations, largely for business and internal processes (not for competitive or proprietary manufacturing). In today's world, improved from "industrial tourism" to research, largely through participation in online databases.

- *Brainstorming.* The popular technique of obtaining group ideas as to cause (for improvement) or as to features (for planning).

- *Carryover analyses.* Usually a matrix that depicts the degree of carryover of design elements, with particular regard to failure proneness.

- *Competitive analyses.* Usually a matrix depicting a feature-by-feature comparison to the competition, with particular regard to "best-in-class" targets.

- *Control chart.* Widely used depiction of process change over time. The most popular is the Shewhart control chart for averages.

- *Criticality analyses.* Usually a matrix that depicts the degree of failure of a feature or component against the ranking of customer needs, along with responsibilities detailed for correction.

- *Data collection: focus group.* The popular technique of placing customers in a setting led by a trained facilitator to probe for the understanding of needs.

- *Data collection: market research.* Any of a variety of techniques aimed at answering the three fundamental questions of: What is important to the users? What is the order of the "importances?", and How well do we do in meeting them in that order, as compared to competition?

- *Data collection: surveys.* The passive technique of eliciting answers to preset questions about satisfaction or needs. Usually "closed-ended," with meager space for comments or answers to open-ended questions. Poor return rates are a hallmark of this technique, along with the suspicion that those with dissatisfactions respond at higher rates.

- *Failure mode and effect analyses.* Otherwise called FMEA, the matrix presentation of probability of failure, significance of the failure, and the ease of detection, resulting in a "risk priority number." Higher RPNs are attacked first. Used in both improvement and planning settings, though the chief use is as a design tool.

- *Fault tree analyses.* A graphical presentation of the modes of failure showing events that must occur together ("and") or separately ("or") in order to have the failure occur. Usually shown vertically, with the "and-ed" and "or-ed" events cascading like branches on a tree.

- *Flow diagram.* Extremely popular depiction of a process, using standard symbols for activities and flow directions. Originated in software design during the 1950s and evolved into the process mapping widely used today.
- *Glossary.* The chief weapon used to remove the ambiguity of words and terms between customers and providers. A working dictionary of in-context usage, such as, What does "comfortable" mean for an office chair?
- *Histogram.* Widely used bar graphs that show the frequency of observation for continuous data collected and sorted into ranges called *cells.* It is characteristic of all histograms in that they show shape, centering, and width, and that these properties can be used to investigate further.
- *Pareto analyses.* First attributed by J.M. Juran to be widely applicable in sorting out the vital few from the useful many. Named after a nineteenth century economist, the Pareto chart that is usually created shows that roughly 80 percent of the effect is concentrated in about 20 percent of the contributors.
- *Planning network.* A tree diagram depicting the events that occur either in parallel or sequentially when planning anything. Usually shown with the total time needed to complete the event, along with earliest start and subsequent stop dates. Used to manage a particularly complex planning effort. Like techniques include program evaluation and review technique (PERT) and critical path method (CPM). Today's spreadsheet-like project management software usually combines the key features of each.
- *Process analysis technique.* A process flow charting technique that also shows the time necessary to do each task, the dependencies the task requires (such as access to the computer network), and the time "wasted" in between tasks. Usually interview-driven, and requiring a skilled process expert.

- *Process capability.* The term given to any number of tools, usually statistical in nature, which thereby reveal the ability of a process to repeat itself, and the ability of the process to meet its requirements.

- *Salability analyses.* Another matrix tool used to depict the price willing to be borne, or the cost needed to deliver, a given feature of a product.

- *Scatter diagram.* The graphical technique of plotting one variable against another, to determine co-relationship. A prelude to regression analyses to determine prediction equations.

- *Selection matrix.* A matrix tool showing the choices to be made, ranked according to agreed-upon criteria. Used in both improvement and planning settings.

- *SMART method.* Specific, measurable, agreed-upon, realistic, and time-phased dimensions of goal-setting criteria.

- *Customer needs spreadsheet.* A spreadsheet tool depicting the relationship between customer communities and the statements of need. Needs strongly relating to a wide customer base subsequently rise in priority when considering features. Advanced forms of this spreadsheet and others appear as the "house of quality," or quality function deployment (QFD); see the section in this chapter about Design for Six Sigma.

- *Needs analysis spreadsheet.* A spreadsheet tool used to "decompose" primary statements of need into other levels. Thus, "economical" for a new car purchaser might break down further to purchase price, operating costs, insurance costs, fuel economy, and resale value. Decomposing needs has the principal benefit of single point response and measurement if taken to the most elemental level.

- *Product design spreadsheet.* A continuation of the customer needs spreadsheet, further developing the features and feature goals that map to the customer needs. The features with the strongest relationship to needs are elevated in priority when considering the processes used to make them.

- *Tree diagram.* Any of a variety of diagrams depicting events that are completed in parallel or simultaneously as branches of a tree. Less refined than the planning network, but useful to understand the activities from a "big picture" perspective.

- *Value analysis*. A matrix depiction of customer needs and costs required to support or deliver a given feature to meet that need. A close cousin to the salability analysis.

THE SIX SIGMA MODEL OF QUALITY PLANNING: DESIGN FOR SIX SIGMA

Product design is the creation of a detailed description for a physical good or service, together with the processes to actually produce that good or service. In quality theory terms, product design means establishing quality goals and putting in place the means to reach those goals on a sustained basis. In "Six Sigma" terms, product design (Design for Six Sigma [DFSS]) means contemporaneously creating a design for a product, and also the process to produce it in such a way that defects in the product and the process are not only extremely rare, but also pre-dictable. What is more, defects are rare and predictable, even at the point when full-scale production begins.

To achieve this level of excellence and its attendant low costs and short cycle times, as well as soaring levels of customer satisfaction, requires some enhancements to traditional design methods. For example, each DFSS design project starts with an identification of customers and a detailed analysis and understanding of their needs. Even "redesign" starts at the beginning because all successful designs are based on customer needs, and in this world of rapid change, customer needs—and even customers—have a way of rapidly changing.

Another example is the widespread intensive use of statistical methods in DFSS. The power of the information gained from statistical analyses provides the means to achieve Six Sigma levels of quality, which is measured in parts per million.

DFSS is carried out in a series of phases known as DMADV. DMADV stands for: define-measure-analyze-design-verify.

The discussion that follows does not cover all the details of procedures and tools used in DMADV; that would require many hundreds of pages, and can be found elsewhere in published form. We will, however, attempt to acquaint the reader with what any manager needs to know about the purpose, the issues, the questions, and the sequence of steps associated with the respective phases of DMADV.

DESIGN FOR SIX SIGMA

OVERVIEW

A "new" codification of the process for developing quality products is known as Design for Six Sigma (DFSS). It combines the concept of quality planning with the popular goal of Six Sigma quality. The DFSS process directs the designers of the product to create their designs so that manufacturing can produce them at Six Sigma quality levels. In the case of services, it means developing the service process so that it can be delivered at Six Sigma quality levels.

DFSS is targeted at design activities that result in a new product, a new design of an existing product, or the modification of an existing design. It consists of five phases in the following sequence: Define↔Measure↔Analyze↔Design↔ Verify. Table 3.1 expands on the activities of each phase.

TABLE 3.1   Major Activites in Phases of DFSS

DEFINE	MEASURE	ANALYZE	DESIGN	VERIFY
• Initiate the project. • Scope the project. • Plan and manage the project.	• Discover and prioritize customer needs. • Develop and prioritize CTQs. • Measure baseline performance.	• Develop design alternatives. • Develop high-level design. • Evaluate high-level design.	• Optimize detail-level design parameters. • Evaluate detail-level design. • Plan detail design verification tests. • Verify detail and design of product. • Optimize process performance.	• Execute pilot/ analyze results. • Implement production process. • Transition to owners.

DEFINE PHASE

A project enters the define phase when it is officially launched by the management team. A project gets launched when it meets the selection criteria established to evaluate the entire set of nominated projects. Some typical considerations for selection criteria may be one or more of the following:

- Alignment with business strategy
- Significant contribution to business performance
- Significant opportunity to improve customer satisfaction
- Provides competitive advantage

The define phase establishes the project as a business initiative worthy of completion. A key task in this regard is to create the initial business case that validates the selection rationale and establishes the business justification either through reduced cost (replacing and existing design), increased sales, or an opportunity to enter a new market.

It may be necessary for the management team to work closely with the project design team to refine the design problem. This will lead to a more accurate scope of the project and will assure a common understanding of what is to be modified, redesigned, or created. This should lead to a better understanding of the resources required to execute the project. As research has indicated, projects that fail to deliver the expected results frequently get off track at the start, when the project is being defined.

The management team nominates a leader for the team. Typically, the leader of the project design team is designated the project/program manager. In companies pursuing Six Sigma, they often designate the leader as a black belt or a master black belt. The black belt designation implies the person has been certified to have the requisite skills to lead the project team to successful completion of the project. The black belt designation will be assumed for the remainder of this section.

The black belt meets with the champion—who is the management sponsor with vested interest in the success of the project—to select the core cross-functional team that will be responsible for the activities required to carry out the project. An often overlooked or discounted responsibility of the black belt is project management skills. The body of knowledge for project management covers a wide variety of planning, interpersonal communication, analysis skills, and tools.

MEASURE PHASE

The measure phase in the DMADV sequence establishes what is to be designed. The three activities given in Table 3.1 can be described as getting to know the customer of the product or service, and determining where the company stands in meeting the expectations of those customers.

Getting to know the customer essentially starts with understanding the markets currently served by the company, or the new markets potentially served by a proposed new product. The list of customers may be extensive. The first task is to identify the key customers. This is primarily the task of marketing; however, the project team has the responsibility to verify the market/customer segments identified and to complete the customer needs analysis.

The result of the team's analysis of the marketing information culminates in identification of the critical-to-quality (CTQ) requirements that must be satisfied by the final design. Customer needs are evaluated and translated into measurable characteristics that can be verified. CTQs are stated in specific technical terms that relate to the organization and become the measurable goals (specifications) for product performance and ultimate success.

The project team may use several means to set the goals for each CTQ. Some tools include competitive benchmarking, competitive analysis, and stretch objectives for current performance. The result is a combination of customer stated requirements, and requirements that may not be generally addressed or known by the customer. In addition to these tools, design teams are now exploring the use of the I-TRIZ Innovative Problem-Solving methodology to look at the design issues in new and innovative ways that may lead to a real breakthrough.

The measure phase ends with the assessment of the current baseline performance of the enumerated CTQs and performance of risk assessments. To establish these baselines, typical process capability methods and tools are utilized. These include:

1. Establish the ability of the measurement system to collect accurate data using measurement system analysis (MSA)
2. Measure the stability of the process using statistical process control techniques
3. Calculate the capability and sigma level of the process

To evaluate risk, the team may make use of tools such as product failure mode effects analysis (PFMEA), process failure mode effects analysis (FMEA), and the I-TRIZ tool anticipatory failure determination (AFD).

Another tool employed by some design project teams is the quality function deployment (QFD) matrix. The matrix has several components that are basically laid out: 1) horizontally for customer information (needs, weighting factors, customer assessments, and competitive evaluations), and 2) vertically for technical information (technical requirements, competitive technical evaluations, the operation goals, and weighting factors). The intersection of the horizontal and vertical sections is an area to document relationships between the customer needs and the technical responses to meet those needs.

The QFD Matrix (or simpler versions) is meant to highlight the strengths and weaknesses that currently exist. In particular, the weaknesses represent gaps that the design team must shrink or overcome. The demand on the team then, is to provide innovative solutions that will economically satisfy customer needs. Keeping this matrix up-to-date provides a running gap analysis for the team.

ANALYZE PHASE

The main purpose of the analyze phase is to select a high-level design and develop the detailed system design requirements that will be the targets for performance of the detailed design. This is sometimes referred to as system-level design versus the subsystem or component design levels.

The design team develops several high-level alternatives that represent different functional solutions to the collective CTQ requirements. A set of evaluation criteria is then developed, against which the design alternatives will be analyzed. The final configuration selected may be a combination of two or more alternatives. As more design information is developed during the course of the project, the design may be revisited and refined.

After the selection of the high-level design, the team establishes the functional system's architecture. The flow of signals, flow of information, and mechanical linkages indicate the relationship between the subsystems for each design alternative. Hierarchical function diagrams, functional block diagrams, function trees, and signal flow diagrams are commonly used to illustrate these interrelationships. Where possible, models would be developed and simulations run to evaluate the overall system functionality.

The requirements for each subsystem are expressed in terms of their functionality and interfaces. The functionality may be expressed as the system transfer function, which would represent the desired behavior of the system/subsystem. Interfaces are described in terms of the input and output requirements and the controls (feedback, feed-forward, automatic controls). These specifications will be provided to the detail design teams.

In the analyze phase, DMADV analysis tools enable the design team to assess the performance of each design alternative and to test the differences in performance of the competing design alternatives. The results of the these tests lead to the selection of the "best fit design," which is then the basis to move into the next phase, detailed design. These analyses are accomplished using graphical and analytical tools, some of which are:

- Hypothesis testing
- Analysis of variance (ANOVA)

- Box plots
- Multivari charts
- Correlation and regression
- Designed experiments

One of the significant advances affecting this process is the availability of several statistical analysis tools. These software applications, running on desktops or laptop computers, speed up the number crunching required to perform the preceding analysis. This availability has also made it necessary for individuals, who would not normally use these tools, to be trained in the use and interpretation of the results.

DESIGN PHASE

The design phase builds upon the high-level design requirements to deliver an optimized functional design that meets manufacturing and service requirements. Detail designs are carried out on the subsystems and eventually integrated into the complete functional system (product).

DMADV tools focus on optimizing the detail level design parameters. In particular, designed experiments (DOEs) are planned and conducted on the vital few design parameters. DOEs are conducted for several purposes. One purpose may be to determine the best set of features (optimum configuration) to employ. Another purpose may be to obtain a mathematical prediction equation that can be used in subsequent modeling and simulations. Additionally, experiments are designed at differing levels of complexity, from minimal run screening experiments to multi-level replicated designs. Screening experiments typically try to establish which factors influence the system, providing somewhat limited results for modeling. More detailed experiments, including response surface and mixture designs, are conducted to more accurately determine system performance and produce a mathematical equation suitable for prediction and modeling applications.

During the design phase, the design team is also concerned about the processes that must be developed to provide the service or build the product. During the measure phase, the team examines the current capability of the business to deliver the product or service at the expected quality levels (approaching Six Sigma). During the design phase, the team continually updates the design scorecard with the results of DOEs, benchmarking results, process capability studies, and other studies to track the design performance against the established goals. This continues the gap analysis that runs throughout the project.

At the completion of the design phase, the design parameters have been defined and the key parameters optimized. To conclude the design phase requires the goals of the design for performance to be verified through testing of prototype, preproduction models, or initial pilot samples. The design team documents the set of tests, experiments, simulations, and pilot builds required to verify the product/service performance in a design verification test (DVT) plan. Upon completion of the several iterations that occur during the DVT and pilot runs, the design is solidified and the results of testing are summarized.

A design review meeting marks the conclusion of the design phase, when the results of the DVT are reviewed. The design scorecard is updated and each area of the development plan (quality plan, procurement plan, manufacturing plan, etc.) is adjusted as necessary.

VERIFY PHASE

The purpose of the verify phase in the DMADV sequence is to ensure that the new design can be manufactured and field supported within the required quality, reliability, and cost parameters. Following DVT, a ramp-up to full-scale production is accomplished via the manufacturing verification test (MVT). The objective of this series of tests is to uncover any potential production or support issues or problems.

The manufacturing process is typically exercised through one or more pilot runs. During these runs, appropriate process evaluations occur, such as capability analyses and measurement systems analyses. Process controls are verified and adjustments are made to the appropriate standard operating procedures, inspection procedures, process sheets, and other process documentation. These formal documents are handed off to downstream process owners (e.g., manufacturing, logistics, and service). They should outline the required controls and tolerance limits that should be adhered to and maintained by manufacturing and service. These documents come under the stewardship of the company's internal quality systems, many of which are documented around ISO 9000 or other similar standards. One of the considerations of the design team is to assure that the project documentation will conform to the internal requirements of the quality system.

The design team should ensure that appropriate testing in a service and field support environment is accomplished to uncover potential lifetime or serviceability issues. These tests will vary greatly, depending on the product and industry. These tests may be lengthy, and possibly not conclude before production launch. The risks associated with not having completed all tests depend on the effectiveness of earlier testing and the progress of final MVT tests that are underway.

A final design scorecard should be completed and all key findings should be recorded and archived for future reference. The team should complete a final report that includes a look back at the execution of the project. Identifying and discussing the positive and not-so-positive events and issues will help the team learn from any mistakes made and provide the basis for continuing improvement of the DFSS sequence.

DMADV: DEFINE

OVERVIEW

The define phase sets the tone for the entire design project in that it establishes its goals, charter, and infrastructure. During this phase, activities are shared between both the management team and the chartered project design team. Management has the ultimate responsibility to define the design problem: what is to be modified, redesigned, or newly created.

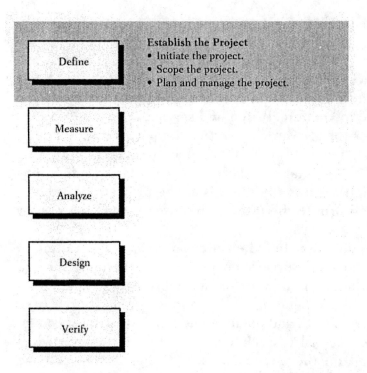

Projects are nominated consistent with the overall business strategy, and selected based on their optimal contribution to that strategy.

A key task in the define phase is to create the initial business case that validates the selection rationale and establishes the business justification either through reduced product cost, increased sales, or entire new market opportunities. The initial business case work is conducted under the auspices of the management team, and then validated and updated continuously by the design team through the subsequent phases of the design project.

The management team nominates a black belt to lead the design project. The champion, who is the management sponsor with vested interest in the success of the design, in conjunction with the black belt, is responsible for selecting a cross-functional team that will conduct all the activities to complete the design and carry it into production.

With the nomination of the multi-functional team, the project is launched. Full responsibility for design success is transferred to the black belt and his/her team. Management participation continues throughout the design effort through the champion's advisory and monitoring role that includes periodic updates from the black belt and team.

The design team, under the black belt's leadership, establishes the project plan that includes resource allocation and a task list via a project timeline. The project plan is based on a leveraged design strategy developed by the team while utilizing the I-TRIZ tool directed evolution (DE).

In summary, the key deliverables that are required to complete the define phase are:

DEFINE: DELIVERABLES

- Initial business case is developed
- Design strategy developed with directed evolution (DE)
- Design project is established; leadership and team is selected
- Project charter is drafted, including project mission statement and design objectives
- Project plan is created

Questions to Be Answered

- What are the design goals or objectives of the project?
- What is the specific mission of the project team?
- What is the business case that justifies the project?
- What charter will the team receive from management empowering them to carry out the project?
- Who will be the black belt, the champion, the team members?
- What will be the project strategy or plan?
- How will the project be managed?

INITIATE THE PROJECT

- Identify service/product concept
- Develop business case

- Define project leadership
- Draft charter

SCOPE THE PROJECT

- Define project scope
- Commit resources
- Kickoff project
  Champion
  Black belt
  Cross-functional team
  Design budget
  Establish working relationships

PLAN AND MANAGE THE PROJECT

- Determine project management approach
  Project controls
  Project reviews
  Communication plan
  Issues and risk management
- Develop project plan
  Planning requirements
  Define deliverables/timeline
- Develop organizational change management approach

DMADV: MEASURE

OVERVIEW

The measure phase in the DMADV sequence is mainly concerned with identifying the key customers, determining what their critical needs are, and what are the measurable CTQs necessary for a successfully designed product.

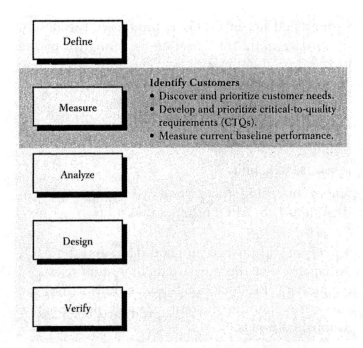

Define

Measure

Identify Customers
- Discover and prioritize customer needs.
- Develop and prioritize critical-to-quality requirements (CTQs).
- Measure current baseline performance.

Analyze

Design

Verify

An initial assessment of our markets and customer segmentation by various factors is required to identify the key customers. This assessment is normally completed by the marketing organization and is then reviewed and verified by the design team. However, it is the design team's responsibility to complete the customer needs analysis and compile its results into a prioritized tabulation of customer needs.

The design team transforms the critical customer needs into measurable terms from a design perspective. These translated needs become the CTQs that must be satisfied by the design solution. Competitive benchmarking and creative internal development are two additional sources to generate CTQs. These areas probe into design requirements that are not generally addressed or possibly even known by the customer. In particular, the I-TRIZ product innovative problem solving (IPS) will assist the design team in a more accurate and complete description of the design problem. The result is a set of consistently defined and documented design CTQs.

Once the prioritized list of CTQs is produced, the design team proceeds to determine the baseline performance of the existing product and production process. The current baseline performance is determined in terms of four components:

- Measurement systems analysis (MSA)
- Product capability
- Production process capability
- Risk assessment by using tools such as Product FMEA, Process FMEA, and IPS, AFD: product and process

Six Sigma decisions are based on good data, and the MSA is a method to ensure that the measurement system uses and yields high-quality data. Product capability analysis assists the design team in assessing the current product performance against the established CTQs.

Finally, a design scorecard is created that tracks the design evolution toward a Six Sigma product performance. This tool is used in the attempt to predict what the final product defect level will be after integration of all the design elements.

In summary, the key deliverables that are required to complete the measure phase are:

- Prioritized list of customer needs
- Prioritized list of CTQs
- Current baseline performance:
  MSA
  Product capability
  Production process capability, supported by process flow diagram
  Product and process risk assessment
- Initial design score card

MEASURE: DELIVERABLES

- Prioritized list of customer needs
- Prioritized list of CTQs
- Current baseline performance:
  Measurement system analysis
  Product capability
  Production process capability, supported by process flow diagram
  Product and process risk assessment
- Initial design scorecard

QUESTIONS TO BE ANSWERED

- What customer needs must the new product meet?
- What are the critical product and process requirements that will enable the customer needs to be met?
- How capable is our current product and production process in meeting these requirements?
- How capable must any new product and production process be in order to meet these requirements?

DISCOVER AND PRIORITIZE CUSTOMER NEEDS

- Identify customers
  Assess markets and customer segments
  Analyze flow diagram of any existing production process
- Plan to collect customer needs
  From internal customers
  From external customers
- Collect list of customers' needs in their language
- Translate, analyze, and prioritize customers' needs
- Establish units of measure and sensors

DEVELOP AND PRIORITIZE CRITICAL-TO-QUALITY REQUIREMENTS

- Identify product/process requirements necessary to meet each translated customer need
- Confirm how this CTQ is to be measured (unit of measure and sensors)
- Determine target value for each CTQ
- Determine upper and lower specification limits for each CTQ
- Establish target permissible defect rate (DPMO, Sigma) for each CTQ

MEASURE CURRENT BASELINE PERFORMANCE

- Analyze capability of the measurement system
- Determine product capability
- Determine production process capability, supported by process flow diagram
- Perform product and process risk assessment
- Create initial design scorecard

DMADV: ANALYZE

OVERVIEW

The main purpose of the analyze phase in the DMADV sequence is to select a high-level design from several design alternatives and develop the detailed design requirements against which a detailed design will be optimized in the subsequent design phase.

The starting point in the high-level design is to perform a functional analysis of the CTQs established in the measure phase that results in a high-level functional design. This should satisfy the CTQs in an economically viable manner.

The design team develops several high-level design alternatives that represent different functional solutions to the stated

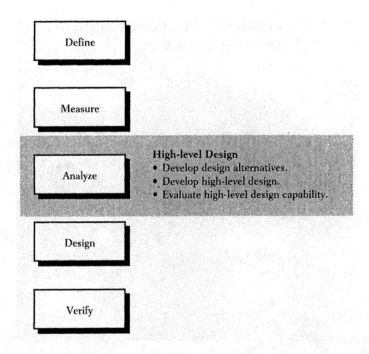

Define

Measure

Analyze

High-level Design
* Develop design alternatives.
* Develop high-level design.
* Evaluate high-level design capability.

Design

Verify

functional design requirement. These alternatives are analyzed against a set of evaluation criteria and one of them, or combination of alternatives, is selected to carry forward as the preferred "high-level design." The development of alternatives and the selection of the preferred alternative are both iterative processes: as progressively more design information is developed, the iterative nature inherent in design requires that several passes be made to ensure that the most capable high-level design is carried forward.

Several of the tools utilized by the team in the measure phase to establish baseline performance are applied again in an attempt to predict the performance of the high-level design against the CTQs. Process capability studies, product functionality, and capability analysis, risk analysis, and financial analysis are the analytical instruments used to predict performance.

After the evaluation is completed and the best high-level design is validated, the design team performs a final analysis that results in a "best fit design." This design furnishes the basis to develop the detailed design requirements that serve as input to the detailed design phase.

In summary, the key deliverables that are required to complete the analyze/high-level design phase are:

- Design alternatives
- Selected high-level design
- Results of high-level design capability/risk analysis
- Best-fit design
- Detailed design requirements

ANALYZE: DELIVERABLES

Develop a high-level product or service and process design and detail design requirements.

- Design alternatives are developed
- Best alternative is selected
- High-level design elements are developed to best-fit design
- High-level design capability is predicted
- Detail design requirements are defined
- Key sourcing decisions are made

QUESTIONS TO BE ANSWERED

- What design alternatives could be employed in the new product/ process service?
- Which is the "best" alternative?
- What are the requirements for the detailed design?

DEVELOP DESIGN ALTERNATIVES

- Perform functional analysis
  Define product/service functions
  Deploy/allocate CTQs to functions
- Develop design alternative

Develop design elements

Benchmark best performance and processes

Apply appropriate technology

Perform capability analysis

Develop initial build/production plan

Establish initial business cause

- Select optimum alternative(s)

Establish selection criteria

Evaluate against criteria (CTQs)

DEVELOP HIGH-LEVEL DESIGN

- Develop high-level design elements

Process, information, facilities, human, materials, equipment, etc.

- Develop quantitative requirements for design elements

Determine capacities and operating ranges

Deploy functional/process requirements to detailed design requirements

- Identify critical product introduction resources

Design resources

Production resources

Other (systems, logistics, distribution, marketing, outsourcing, etc.)

- Document high-level design

EVALUATE HIGH-LEVEL DESIGN CAPABILITY

- Assess capability of high-level design

Define evaluation criteria

Perform qualitative evaluation (capability analysis/gap analysis)

Generate functional options for remaining gaps

- Determine best-fit design

  Integrate product/service functions into best-fit configuration

- Obtain customer feedback

  High-level design review with customer/stakeholder

- Develop initial test plan, service plan, and EIA
- Perform risk assessment

  Develop exit strategies

- Finalize detailed design requirements

DMADV: DESIGN

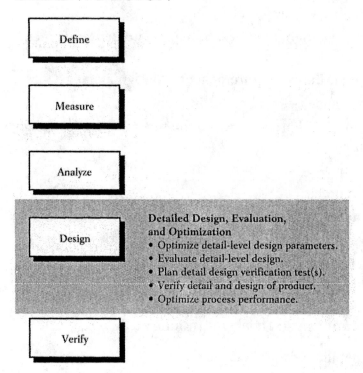

OVERVIEW

The design phase of the DMADV sequence builds upon the detailed design requirements to deliver an optimum detailed functional design that also meets the manufacturing and service requirements.

Using the vital few design parameters determined in the previous analyze/high-level design phase, designed experiments (DOEs) are conducted to optimize the detail design around key design parameters. These result in an optimum detailed parametric design represented by a mathematical prediction equation. Most of the experimentation assumes linear relationships between the design parameters and overall performance. However, tools such as response surface methods are employed to handle non-linear models.

These designed experiments provide the inputs for the tolerance design that concentrates on the selective reduction of tolerances to reduce variation and quality loss.

Using appropriate design methods and tools (DFM/DFA, reliability, and serviceability analysis), the design team examines the capabilities of the current production process and related systems against the new design. A final risk analysis, utilizing traditional tools and the I-TRIZ tool AFD is conducted. Based on these analyses, a final design is developed to match the projected operational (manufacturing and service) capabilities.

The design phase is concluded with the design team conducting a design verification test (DVT) that validates the detail design by utilizing such tools as simulation, prototyping, and pilot testing. The results of the DVT are summarized and presented in a formal design review.

The design scorecard is updated again with the final design information and the latest DVT results.

In summary, the key deliverables that are required to complete the design phase are:

- Optimized design parameters (nominal values that are most robust)
- Prediction model
- Optimal tolerances and design settings
- Detailed functional design

- Reliability/lifetime analysis results
- DVT results
- Updated design scorecard

DESIGN: DELIVERABLES

- Optimized design parameters (elements)—nominal values that are most robust
- Prediction model
- Optimal tolerances and design settings
- Detailed functional design
- Reliability/lifetime analysis results
- Design verification test results
- Updated design score card

QUESTIONS TO BE ANSWERED

- What detailed product design parameters minimize variation in product performance?
- How do we assure optimum product reliability?
- How do we assure simplicity and ease of manufacture?
- What detailed process parameters consistently and predictably minimize production process variation around target values?

OPTIMIZE DETAIL-LEVEL DESIGN PARAMETERS

- Develop detail-level CTQs
- Develop detail-level design features
- Conduct fractional factorial experiments
- Conduct 2k and, if needed, full factorial experiments
- Document product design variables
- Document process variable

Evaluate Detail-Level Design

- Determine tolerances, which reduce variation and quality loss
- Determine economic impact of predictable variation

Plan Detail-Design Verification Test(s) Verify Detail and Design of Product

- Establish comprehensive test plan
- Verify design via DVT
- Assess reliability and availability

Optimize Process Performance

- Update process effectiveness/efficiency measures (Sigma, Cps, and Cpks)
- Revise design scorecard

DMADV: VERIFY

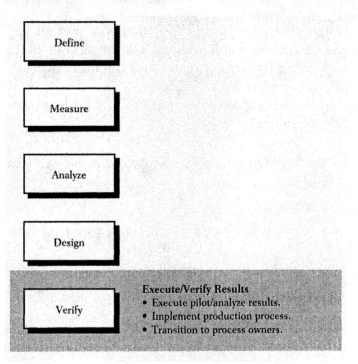

Define

Measure

Analyze

Design

Verify

Execute/Verify Results
- Execute pilot/analyze results.
- Implement production process.
- Transition to process owners.

OVERVIEW

The purpose of the verify phase in the DMADV sequence is to ensure that the new design can be manufactured and field supported within the required quality, reliability, and cost parameters. Upon completion of the several iterations that occur during the DVT and pilot runs, the design is solidified and a ramp-up to full-scale production is accomplished via the manufacturing verification test (MVT) to highlight any potential production issues or problems.

The design team should ensure that appropriate testing in a service and field support environment is accomplished to uncover potential lifetime or serviceability issues. These tests will vary greatly, depending on the product and industry.

A key task is to record all design documents and process controls plans (including guidelines for self-control) into a robust set of standard operating procedures. These formal documents are handed off to downstream process owners (e.g., manufacturing, logistics, and service). They should outline the required controls and tolerance limits that should be adhered to and maintained by manufacturing and service.

A final design scorecard should be completed and all key findings should be recorded and archived for future reference. The team should complete a final report and celebrate the successful completion of a Six Sigma design.

In summary, the key deliverables that are required to complete the verify phase are:

- MVT results (including pilot-scale production processes and scale-up decision)
- Transition documents
- Control plans (including plans for self-control and mistake proofing)
- Final design scorecard
- Final project report (including established audit plan)

Verify: Deliverables

Verify product/process performance against project targets.

- Pilot build is complete
- Pilot tests are completed/results are analyzed
- Scale-up decision(s) are made
- Full-scale processes are built and implemented
- Business results are determined/analyzed
- Processes are transitioned to owners
- DFSS project is closed

Questions to Be Answered

- Is the product/process meeting the specifications and requirements?
- Is the production process "owned" by the business?

Execute Pilot/Analyze Results

- Build pilot-scale processes
  Limited scale, full functionally
- Pilot testing and evaluation
  Pilot results/redesign
  Scale-up decision
- Implementation planning
  Full-scale build plan
  Transition plan
  Documentation

Implement Production Process

- Build full-scale processes
- Start-up and testing
  CAP

Training

Commercialization

- Performance evaluation

Full-scale results (yield, DPMO, demand stability)

TRANSITION TO PROCESS OWNERS

- Turnover to operations and plant maintenance
- Transition to process management

Control processes

Further commercialization

- DFSS project closure

INFRASTRUCTURE, ROLES, AND TRAINING

(Please consult the final pages of Chapter 8, "Breakthroughs in Current Performance.")

One difference between preparing an organization for DFSS, as opposed to DMAIC, is that unlike DMAIC, there is little need for everyone to earn a belt in DFSS. The DFSS subject matter is very extensive, and some of it is quite sophisticated and complex. It is satisfactory to restrict DFSS training to engineers, designers, interested managers, and technicians—all those who would lead or be members of a DFSS design project team.

HIGH POINTS OF "THE PLANNING PROCESSES"

- Planning is the first element of the Juran Trilogy
- The planning process creates designs of products (goods, services, information) together with processes—including controls—to produce the products

- The planning processes can be used to design or redesign. The classic "Juran" model consists of the following basic steps:

  Establish a project and its design goals

  Identify the customers

  Discover the customers' needs

  Develop the product features (to meet the customer needs)

  Develop the process features (to produce the product features)

  Develop process controls and transfer to operations

- The Six Sigma version of the classic model consists of the following same basic steps, with different labels:

  Define (establish the project)

  Measure (identify customers and their needs, including CTQs—"critical to quality characteristics"—to establish required product features)

  Analyze (high-level design incorporating CTQs)

  Design (detailed design, evaluation, and optimization)

  Verify (execute/verify results)

THE CONTROL
PROCESSES

This chapter* presents the second element of the trilogy: Control—a managerial process assuring that products and processes perform to standard by either *preventing* or *correcting* undesirable, unwanted, "bad" change in performance. Control seeks to provide stability by continuously assuring that standards are met. In finance, one speaks of "financial control," which consists of measuring actual expenditures, comparing expenditures to budget, and taking action on overspending.

In other operations, one speaks of "quality control," which consists of measuring actual performance, comparing actual to desired performance (the standard, the specification, etc.), and taking corrective action on bad differences. Control takes many forms (hence the chapter title, "Control Processes"). The various forms are nevertheless all based on the elements of the feedback loop, a feature they have in common. Control occurs in many places in an organization and at all levels. It utilizes many tools and techniques—some quite sophisticated, some quite simple.

Control is an integral part of the other two basic managerial processes: planning and breakthrough. In planning, controls

*The authors wish to acknowledge J. M. Juran and A. Blanton Godfrey. This chapter is based largely on and adapted from Section 4 ("The Quality Control Process") of *Juran's Quality Handbook*, 5th edition, 1999, which they authored.

are built into designs to assure that the planned standards, specifications, product features, process features, etc., actually are produced in practice, under operating conditions. In breakthrough, controls are built into the various types of improvements created by the various types of breakthrough to assure that the original solutions by each breakthrough are sustained and problems do not return or reoccur.

From a managerial point of view, self-control is of special interest. If management provides employees all the elements of self-control, then they will have practically all the means necessary to be successful on the job and to keep a process in a state of control. The only significant limitation is human physiology and psychology, which accounts for most of the errors people make, even when in a state of self-control. The effect of these limitations can be minimized by appropriate mistake-proofing tools.

In planning, an essential step in designing any product, service, or process is assuring that the people who must produce the product and operate the process are in a state of self-control. Similarly, in breakthrough, a necessary feature of the design of any improvement or remedy is assuring self-control on the part of those who will implement the remedial changes.

In the authors' experience, numerous performance/quality problems stem from a lack of self-control. Consequently, an effective (and often relatively quick) method to solve many performance problems is to conduct a self-control assessment, identify the gaps (the elements, or sub-elements, of self-control that a given individual is lacking), and fill the gaps. Self-control is described further in this chapter.

Figure 4.1 describes the relationships between the fundamental managerial processes. It is important to concentrate on the two "zones of control." In Figure 4.1, we can easily see that although the process is in control in the middle of the chart, we are running the process at an unacceptable level of waste. What is necessary here is not more control, but improvement actions to change the level of performance.

After the improvements have been made, a new level of performance has been achieved. Now it is important to establish new controls at this level to prevent the performance level from

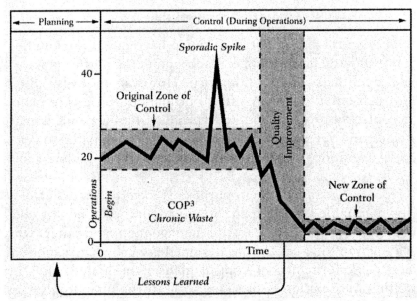

FIGURE 4.1   The Juran Trilogy diagram. (Juran Institute, Inc., Wilton, CT)

deteriorating to the previous level, or even worse. This is indicated by the second zone of control.

In Figure 4.2, the input for the control process is operating process features developed to produce the product features required to meet customer needs. The output consists of a system of product and process controls, which can provide stability to the operating process.

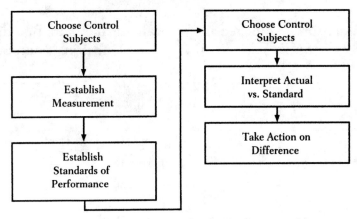

FIGURE 4.2   The input-output diagram for the quality control process.

The Relation to Quality Assurance

Quality control and quality assurance have much in common. Each evaluates performance. Each compares performance to goals. Each acts on the difference. However, they also differ from each other. Quality control's primary purpose is to maintain control of process inputs and outputs. Performance is evaluated during operations, and performance is compared to goals during operations. The resulting information is received and used by the operating forces.

Quality assurance's main purpose is to verify that control is being maintained. Essentially, quality assurance is quality control of control. It seeks to assure management, and others who need to know, that controls as designed are being carried out in practice. Performance of control plans is evaluated, and the resulting information is provided to both the operating forces and others who need to know. Others may include plant, functional, or senior management; corporate staffs; regulatory bodies; customers; and the general public. ISO 9000 is an example of a quality assurance system.

The Feedback Loop

Quality control takes place by use of the feedback loop. A generic form of the feedback loop is shown in Figure 4.3.

The progression of steps in Figure 4.3 is as follows:

1. A sensor is "plugged in" to evaluate the actual quality of the control subject—the product feature (output) or

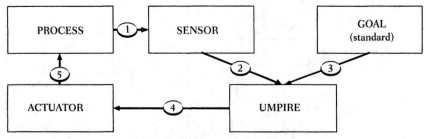

FIGURE 4.3 The generic feedback loop. ("Making Quality Happen," Juran Institute, Inc., senior executive workshop, p. F-3, Wilton, CT)

process feature (input) in question. The performance of a process may be determined directly by evaluation of the process feature, or indirectly by evaluation of the product feature—the product "tells" on the process.

2. The sensor reports the performance to an umpire.

3. The umpire also receives information on what is the quality goal or standard.

4. The umpire compares actual performance to standard. If the difference is too great, the umpire energizes an actuator.

5. The actuator stimulates the process (whether human or technological) to change the performance, so as to bring quality into line with the quality goal.

6. The process responds by restoring conformance.

Note that in Figure 4.3 the elements of the feedback loop are functions. These functions are universal for all applications, but responsibility for carrying out these functions can vary widely. Much control is carried out through automated feedback loops. No human beings are involved. Common examples are the thermostat used to control temperature and the cruise control used in automobiles to control speed.

Another frequent form of control is self-control carried out by a human being. An example of such self-control is the village artisan who performs every one of the steps of the feedback loop. The artisan chooses the control subjects, sets the quality goals, senses what is the actual quality performance, judges conformance, and becomes the actuator in the event of nonconformance.

This concept of self-control is illustrated in Figure 4.4. The essential elements here are the need for the employee or work force team to know what they are expected to do, to know how they are actually doing, and to have the means to adjust their performance. This implies that they have a capable process and have the tools, skills, and knowledge necessary to make the adjustments and the authority to do so.

A further common form of feedback loop involves office clerks or factory workers whose work is reviewed by umpires in

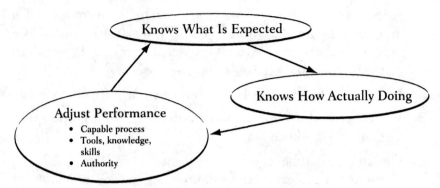

FIGURE 4.4   Self-control. ("Quality Control," Leadership for the Quality
    Century, Juran Institute, Inc., senior executive workshop, p. 5, Wilton, CT)

the form of inspectors. This design of a feedback loop is largely the result of the Taylor system of separating planning from
execution. The Taylor system emerged a century ago, and contributed greatly to increasing productivity. However, Taylor's
effect on quality control was negative. Formal control processes emerged as a corrective reaction to the Taylor system.

THE ELEMENTS OF THE
FEEDBACK LOOP

The feedback loop is a universal. It is fundamental to any problem in control. It applies to all types of operations, whether in
service industries or manufacturing industries, whether for profit or not. It applies to all levels in the hierarchy, from the chief
executive officer to the work force, inclusive. However, there is
wide variation in the nature of the elements of the feedback loop.

In Figure 4.5, a simple flowchart is shown describing the
control process, with the simple universal feedback loop
embedded.

CHOOSE THE CONTROL SUBJECT

Each feature of the product (goods and services) or process
becomes a control subject—a center around which the feedback loop is built. The critical first step is to choose the con-

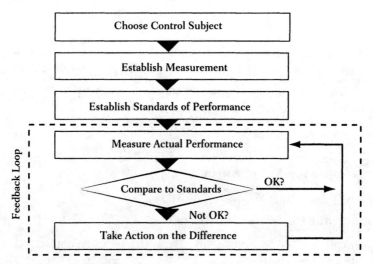

FIGURE 4.5 The quality control process. ("Quality Control," Leadership for the Quality Century, Juran Institute, Inc.)

trol subject. Control subjects are derived from multiple sources, which include:

- Stated customer needs for product features
- Technological analysis to translate customer needs into product and process features
- Process features (inputs), which directly impact the product features (outputs)
- Industry and government standards
- Needs to protect human safety and the environment
- Needs to avoid side effects such as irritations to employees or offense to the neighboring community

At the worker level, control subjects consist mainly of product features and process features set out in specifications and procedures manuals. At managerial levels, the control subjects are broader and increasingly business-oriented. Emphasis shifts to customer needs and to competition in the marketplace. This shift in emphasis then demands added, broader control subjects which, in turn, have an influence on the remaining steps of the feedback loop.

ESTABLISH MEASUREMENT

After choosing the control subject, the next step is to establish the means of measuring the actual performance of the process or the quality level of the goods or services. In establishing the measurement, we need to clearly specify:

- The means of measurement (the sensor)
- The frequency of measurement
- The way the data will be recorded
- The format for reporting the data
- The analysis to be made on the data to convert the data to usable information
- Who will make the measurement

ESTABLISH STANDARDS OF PERFORMANCE: PRODUCT GOALS AND PROCESS GOALS

For each control subject, it is necessary to establish a standard of performance—a quality goal (also called targets, objectives, etc.). A standard of performance is an aimed-at achievement toward which effort is expended. Table 4.1 gives some examples of control subjects and the associated goals.

The prime goal for products is to meet customer needs. Industrial customers often specify their needs with some degree of precision. Such specified needs then become quality goals for the producing company. In contrast, consumers tend to state their needs in vague terms. Such statements must then be translated into the language of the producer in order to become product goals. (Please see Chapter 3, "The Planning Processes.")

Other goals for *products*, which are also important, are reliability and durability. Whether the products meet these goals can have a critical impact on customer satisfaction and loyalty, and on overall costs. The failures of products under warranty can seriously impact the profitability of a company through both direct costs and indirect costs (loss of repeat sales, word of mouth, etc.).

The *processes*, which produce products, have two sets of quality goals:

1. To produce products that meet customer needs. Ideally, each and every unit of product should meet customer needs.

2. To operate in a stable and predictable manner. In the dialect of the quality specialist, each process should be "under control." This will be described further in the section titled Process Conformance. These goals may be directly related to the costs of producing the goods or services.

Quality goals may also be established for departments or persons. Performance against such goals then becomes an input to the company's reward system. Ideally, such goals should be:

- *Legitimate.* They should have undoubted official status.
- *Measurable.* So that they can be communicated with precision.
- *Attainable.* As evidenced by the fact that they have already been attained by others.
- *Equitable.* Attainability should be reasonably alike for individuals with comparable responsibilities.

TABLE 4.1   Examples of Control Subjects and Associated Quality Goals

CONTROL SUBJECT	GOAL
Vehicle mileage	Minimum of 25 mi/gal highway driving
Overnight delivery	99.5% delivered prior to 10:30 a.m. next morning
Reliability	Fewer than three failures in 25 years of service
Temperature	Minimum 505°F; maximum 515°F
Purchase order error rate	No more than three errors/1000 purchase orders
Competitive performance	Equal or better than top three competitors on six factors
Customer satisfaction	90% or better rate; service outstanding or excellent
Customer retention	95% retention of key customers from year to year
Customer loyalty	100% of market share of over 80% of customers

Quality goals (standards) may be set from a combination of the following bases:

- Goals (standards) for product features and process features are largely based on technological analysis.
- Goals for departments and persons should be based on benchmarking, rather than historical performance.

Quality goals at the highest levels are evolving. The emerging practice is to establish goals on matters such as meeting customers' changing needs, meeting competition, maintaining a high rate of quality improvement, improving the effectiveness of business processes, and revising the planning process to avoid creating new failure-prone products and processes.

MEASURE ACTUAL PERFORMANCE

A critical step in control is to measure the actual performance of the product or the process. To make this measurement, we need a sensor, a device to make the actual measurement.

THE SENSOR

A "sensor" is a specialized detecting device. It is designed to recognize the presence and intensity of certain phenomena, and to convert the resulting data into "information." This information then becomes the basis of decision-making. At lower levels of organization, the information is often on a real-time basis and is used for current control. At higher levels, the information is summarized in various ways to provide broader measures, detect trends, and identify the vital few problems.

The wide variety of control subjects requires a wide variety of sensors. A major category is the numerous technological instruments used to measure product features and process features. Familiar examples are thermometers, clocks, yardsticks, and weight scales. Another major category of sensors is the data systems and associated reports that supply summarized information to the managerial hierarchy. Yet another category involves the use of human beings as sensors. Questionnaires and interviews are also forms of sensors.

Sensing for control is done on a huge scale. This has led to the use of computers to aid in the sensing and in conversion of the resulting data into information.

Most sensors provide their evaluations in terms of a unit of measure—a defined amount of some quality feature—which permits evaluation of that feature in numbers. Familiar examples of units of measure are degrees of temperature, hours, inches, and tons. A considerable amount of sensing is done by human beings. Such sensing is subject to numerous sources of error.

COMPARE TO STANDARDS

The act of comparing to standards is often seen as the role of an umpire. The umpire may be a human being or a technological device. Either way, the umpire may be called on to carry out any or all of the following activities:

1. Compare the actual quality performance to the quality goal.
2. Interpret the observed difference; determine if there is conformance to the goal.
3. Decide on the action to be taken.
4. Stimulate corrective action.

These activities require elaboration and will shortly be examined more closely.

TAKE ACTION ON THE DIFFERENCE

In any well-functioning quality control system, we need a means of taking action on the difference between desired standards of performance and actual performance. We need an actuator. This device (human, technological, or both) is the means for stimulating action to restore conformance. At the worker level, it may be a keyboard for giving orders to an office computer or a calibrated knob for adjusting a machine tool. At the management level, it may be a memorandum to subordinates.

THE PROCESS

In all of the preceding discussion, we have assumed a process. This may also be human, technological, or both. It is the means for producing the product features, by means of process features, each of which is a potential control subject. All work is done by a process that consists of an input, labor, technology, procedures, energy, materials, and output.

THE PDCA CYCLE

There are many ways of dividing the feedback loop into elements and steps. Some of them employ more than six elements; others employ fewer than six. A popular example of the latter is the so-called PDCA cycle, as shown in Figure 4.6. Deming (1986) referred to this as the Shewhart cycle, which is the name many still use when describing this version of the feedback loop.

In this example, the feedback loop is divided into four steps labeled Plan, Do, Check, and Act. These steps correspond roughly to the six steps described previously:

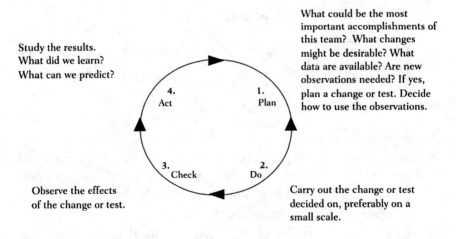

Step 5. Repeat Step 1, with knowledge accumulated.
Step 6. Repeat Step 2, and onward.

FIGURE 4.6   The PDCA cycle.

- "Plan" includes choosing control subjects and setting goals.
- "Do" includes running the process.
- "Check" includes sensing and umpiring.
- "Act" includes stimulating the actuator to take corrective action.

THE PYRAMID OF CONTROL

Control subjects run to large numbers, but the number of "things" to be controlled is far larger. These things include the published catalogs and price lists sent out, multiplied by the number of items in each; the sales made, multiplied by the number of items in each sale; the units of product produced, multiplied by the associated numbers of quality features; and so on for the numbers of items associated with employee relations, supplier relations, cost control, inventory control, product and process development, etc.

A study in one small company employing about 350 people found that there were over a billion things to be controlled.

There is no possibility for upper managers to control huge numbers of control subjects. Instead, they divide up the work of control, using a plan of delegation, somewhat as depicted in Figure 4.7.

This division of work establishes three areas of responsibility for control: control by nonhuman means, control by the work force, and control by the managerial hierarchy.

CONTROL BY NONHUMAN MEANS

At the base of the pyramid are the automated feedback loops and error-proofed processes, which operate with no human intervention other than maintenance of facilities (which, however, is critical). These nonhuman methods provide control over the great majority of things. The control subjects are exclusively technological, and control takes place on a real-time basis.

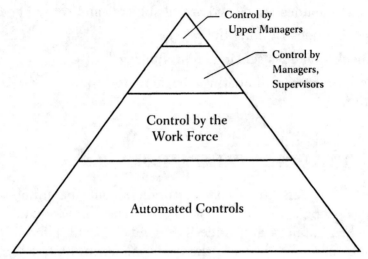

FIGURE 4.7   The pyramid of control. (Making Quality Happen, Juran Institute, Inc., senior executive workshop, p. F-5, Wilton, CT.)

CONTROL BY THE WORK FORCE

Delegating such decisions to the work force yields important benefits in human relations and in conduct of operations. These benefits include shortening the feedback loop; providing the work force with a greater sense of ownership of the operating processes, often referred to as "empowerment"; and liberating supervisors and managers to devote more of their time to planning and breakthrough.

It is feasible to delegate most quality control decisions to the work force. Many organizations already do. However, to delegate process control decisions requires meeting the criteria of "self-control." To delegate product control decisions requires meeting the criteria for "self-inspection." (See the section in this chapter on Self-Control and Self-Inspection, respectively.)

CONTROL BY THE MANAGERIAL HIERARCHY

The peak of the pyramid of control consists of the "vital few" control subjects. These are delegated to the various levels in the managerial hierarchy, including the upper managers. Managers should avoid getting deeply into making decisions on quality control. Instead, they should:

- Make the vital few decisions.
- Provide criteria to distinguish the vital few decisions from the rest. For an example of providing such criteria, see Table 4.3, which focuses on the fitness for use decision.
- Delegate the rest under a decision-making process, which provides the essential tools and training.

The distinction between vital few matters and others originates with the control subjects. Table 4.2 shows how control subjects at two levels—work force and upper management—affect the elements of the feedback loop.

TABLE 4.2   Contrast of Quality Control at Two Levels—Work Force and Upper Management

	AT WORK FORCE LEVELS	AT MANAGERIAL LEVELS
Control goals	Product and process features in specifications and procedures	Business-oriented, product salability, competitiveness
Sensors	Technological	Data systems
Decisions to be made	Conformance or not?	Meet customer needs or not?

Source: "Making Quality Happen," Juran Institute, Inc.

PLANNING FOR QUALITY CONTROL

Planning for control is the activity that provides the system—the concepts, methodology, and tools—through which company personnel can keep the operating processes stable and thereby produce the product features required to meet customer needs. The input-output features of this system were depicted in Figure 4.2.

THE CUSTOMERS AND THEIR NEEDS

The principal customers of quality control systems are the company personnel engaged in control—those who carry out the steps that form the feedback loop. Such personnel require:

· An understanding of customers' quality needs
· A definition of their own role in meeting those needs

However, most of them lack direct contact with customers. Planning for quality control helps to bridge that gap by supplying a translation of what are customers' needs, along with defining responsibility for meeting those needs. In this way, planning for quality control includes providing operating personnel with information on customer needs (whether direct or translated) and definition of the related control responsibilities of the operating personnel. Planning for quality control can run into extensive detail.

WHO PLANS?

Planning for quality control has in the past been assigned variously to:

· Staff planners who also plan the operating processes
· Staff quality specialists
· Multifunctional teams of planners and operating personnel
· Departmental managers and supervisors
· The work force

Planning for quality control of critical processes has traditionally been the responsibility of those who plan the operating process. For noncritical processes, the responsibility was usually assigned to quality specialists from the Quality Department. Their draft plans were then submitted to the operating heads for approval.

Recent trends have been to increase the use of the team concept. The team membership includes the operating forces and may also include suppliers and customers of the operating process. A recent trend has also been to increase participation by the work force. The concept of self-directing work teams has been greatly expanded in recent years, and includes many of these ideas.

QUALITY CONTROL CONCEPTS

The methodologies of quality control are built around various. concepts such as the feedback loop, process capability, self-control, etc. Some of these concepts are of ancient origin; others have evolved in this century. During the discussion of planning for quality control, we will elaborate on some of the more widely used concepts.

THE FLOW DIAGRAM

The usual first step in planning for quality control is to map out the flow of the operating process. The tool for mapping is the "flow diagram." Figure 4.8 is an example of a flow diagram.

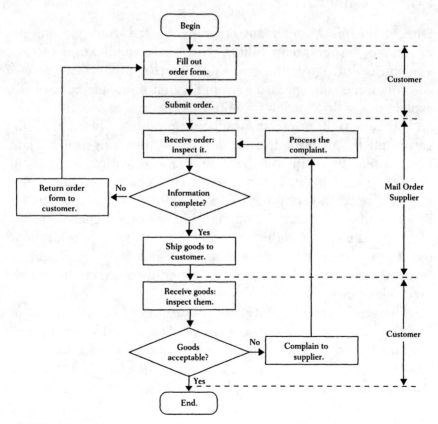

FIGURE 4.8   The flow diagram.

The flow diagram is widely used during planning of quality controls. It helps the planning team to:

- Understand the overall operating process. Each team member is quite knowledgeable about his/her segment of the process, but less so about other segments and about interrelationships.
- Identify inputs and outputs of steps in the flow DIAC, around which the feedback loops are to be built. The nature of these control subjects was discussed previously.
- The control subject.
- Design control stations. (See the following section.)

PROCESS CAPABILITY

One of the most important concepts in the quality planning process is "process capability." The prime application of this concept is during planning of the operating processes.

This same concept also has application in quality control. To explain this, a brief review is in order. All operating processes have an inherent uniformity for producing product. This uniformity can often be quantified, even during the planning stages. The process planners can use the resulting information for making decisions on adequacy of processes, choice of alternative processes, need for revision of processes, and so forth, with respect to the inherent uniformity and its relationship to process goals.

Applied to planning for quality control, the state of process capability becomes a major factor in decisions on frequency of measuring process performance, scheduling maintenance of facilities, etc.

The greater the stability and uniformity of the process, the less the need for frequent measurement and maintenance.

Those who plan for quality control should have a thorough understanding of the concept of process capability and its application to both areas of planning—planning the operating processes, as well as planning the controls.

SEQUENCE OF EVENTS FOR PLANNING A FEEDBACK LOOP FOR CONTROL

The sequence of events for designing and operating a feedback loop for control are as follows:

1. Choose a "control subject," what is to be controlled and measured (e.g., time, temperature, length, diameter, etc.).
2. Choose an appropriate unit of measure (e.g., minutes, degrees centigrade, inches, millimeters, etc.).
3. Establish a standard, the performance target; usually it is a specification at a desired or required level.
4. Establish a sensor, the measuring device (e.g., clock, computer, thermometer, scale, micrometer) and/or the person who takes the measurement—perhaps using or reading the device.
5. Determine how often measurements will be made (e.g., hourly, daily, every piece, etc.), and sample size.
6. Make provision for comparing actual performance to standard, and a criterion for taking action.
7. Decide who/what takes what action: what specific person is authorized to take what specific corrective action to bring the control subject back into conformance with the standard.
8. Take the action.

To illustrate this sequence, consider the following examples:

1. Choose a "control subject," an important input or output characteristic that is to be controlled, such as:
   - Time to answer the telephone or e-mail
   - Time to publish the monthly financial report
   - Number of errors/omissions on a standard form
   - Diameter
   - Cost

- Thickness
- Voltage, etc.

2. Choose an appropriate unit of measure that can describe the control subject, usually in terms of numbers. In the preceding examples, the units of measure would be, respectively:
   - Number of seconds (to answer the phone)
   - Number of days (to publish the monthly financial report)
   - Number of errors on a standard form
   - Thousandths of inches or millimeters (of diameter)
   - Dollars, pounds, etc. (of cost)
   - Inches (thickness)
   - Number of volts (of voltage)

3. Establish a standard, or specification, which is the value in the designed units of a measure that the process should produce, according to the product and process plan. In our example, the standards might be:
   - Less than 30 seconds to answer the phone
   - Ten days maximum to publish the monthly financial report
   - Zero errors on a standard form
   - 0.005 inches plus or minus 0.001 inches in diameter
   - Less than $500 per purchase
   - 1-1/4 inches of thickness
   - 2,500 volts

4. Establish a sensor. A sensor is a device (an automated programmable controller, a computer, a clock, etc.), a person (the operator doing the work being measured), or a person/device team (a person reading a manual gauge or stopwatch). In our example, the sensors might be:
   - A computer recording the time to answer the phone
   - The persons responsible for publishing and those receiving the report note the elapsed time

- A supervisor counting the errors on a standard form
- A programmable controller measuring the diameter
- An accountant reviewing reports of purchases
- An operator reading a manual gauge to determine thickness
- A voltmeter to measure voltage

5. Determine how the measurements will be made and reported (who, what, how often, where in the process, etc.) In our examples:

- The computer might create a weekly summary report of an analysis of times to answer the phones, and automatically e-mail the report to the phone answerers and their supervisors
- The head of finance notes monthly the number of days after the end of the month required to publish each monthly report
- A supervisor will sample completed forms monthly, and share the results with those who fill out the forms, and their supervisors
- The programmable controller perpetually records a run chart of diameters of all products produced. This report can be retrieved, printed out, etc.
- An accountant monthly identifies the number of purchases that exceeded $500, and prepares a summary report that is sent to each purchaser
- The operator measures and records the thickness of each piece he/she produces
- The person running the electrical machinery regularly glances at the voltmeter

6. Make provision for comparing actual performance to the standard. At this point, someone or something makes a comparison of the sensor-generated measure with the standard, and determines the extent to which it meets the standard. If the difference is bad, and it exceeds the stipulated permissible range of measurements that is acceptable, this

triggers the execution of the rest of the feedback loop by signaling the need for corrective action. In our example:

- The phone answerers and their supervisors examine the reports of calls to identify those who took too long to be answered (perhaps the computer software would have already identified and displayed the calls that took too long to answer)

- The head of finance looks at the report of time taken to publish

- The supervisor might analyze patterns of error to discover the most common error types, the persons who make the most errors, etc. Another approach might be to appoint a breakthrough improvement project team to reduce the number of errors on form X

- A supervisor, quality inspector, or perhaps the worker, a quality action team, or a breakthrough improvement team will examine the run charts for patterns and identify defective items that required rework or scrapping

- The accountant glances the list of oversized purchases. If there are any items on the list, the standard is not being met

- The operator notes all items whose recorded measurements are out of specification

- The operator observes and notes voltage readings that are above or below specification

7. The device or person comparing the actual measurements to standard (the "umpire") decides what action to take. This action may be automatic, as in the case of a programmable controller or other device. If a person decides, there may (hopefully) be a specific person or operator designated in the control plan to take a standard action. Sometimes a standard action won't do the job. There may be a wide range of possible actions that could be taken, depending on the specifics of the situation, in which case discretion as to the choice is vested in the person designated by the control plan. In our example:

- The supervisor might meet with each late-answering telephone answerer to seek understanding, generate theories of cause, offer support, offer assistance, etc. A quality action team or a breakthrough improvement project team might be given the mission to reduce the number of late answerers.

- Because so many people and processes are required to publish the end-of-month report, corrective action would likely be to commission a breakthrough improvement team to speed up the monthly reporting process.

- Actions might include more training for error-prone workers, revise the form to make it easier to fill out, transfer hopeless workers to other tasks, assist the worker to pay attention by fool-proofing the process of filling out the form, etc. If a breakthrough improvement team is assigned to reduce errors, after making possible temporary quick fixes, await their discoveries and remedial improvements.

- Possibilities include: adjust the machine that mills the parts, perform designed experiments to seek root causes and ideal methods for operating the milling machine, examine the work habits of successful and unsuccessful workers to discover "the knack" of doing it correctly, etc.

- The accountant could talk to the offending purchaser, review with that person the details of the oversized purchase, and explain that further oversized purchases will result in the overage being deducted from the person's budget. In addition, attention should probably be called to the policy/procedure for requiring purchases over $500 to be approved by someone else. Or, if there is no such policy, decide to create one. A feedback loop is not used for tightening the disciplinary noose to ultimately punish the offender; the feedback loop merely seeks adherence to the standard. (Discipline may at some point be considered, but only as a last resort, after chronic violations.)

- The operator could adjust the voltage output.

8. Take the action. The key point here is that a specific person is designated in the control plan to be responsible for taking the action to close the feedback loop, and adjust the process so it once again produces output to standard.

How does management assure itself that the control plan is actually carried out according to plan? Typically quality assurance is performed in the shape of a periodic, unscheduled internal audit to ascertain that the controls are being correctly applied. The auditor's report goes to upper management.

THE CONTROL SPREADSHEET

The work of the planners is usually summarized on a control spreadsheet. This spreadsheet is a major planning tool. An example is shown in Figure 4.9.

In this spreadsheet, the horizontal rows are the various control subjects. The vertical columns consist of elements of the feedback loop plus other features needed by the operating forces to exercise control so as to meet the quality goals.

Some of the contents of the vertical columns are unique to specific control subjects. However, certain vertical columns apply widely to many control subjects. These include unit of measure, type of sensor, quality goal, frequency of measurement, sample size, criteria for decision making, and responsibility for decision making.

Who Does What?

The feedback loop involves multiple tasks, each of which requires a clear assignment of responsibility. At any control station, there may be multiple people available to perform those tasks. For example, at the work force level, a control station may include setup specialists, operators, maintenance personnel, inspectors, etc. In such cases, it is necessary to agree on who should make which decisions, and who should take which actions. An aid to reaching such agreement is a special spreadsheet similar to Figure 4.9.

PROCESS CONTROL FEATURES / CONTROL SUBJECT	UNIT OF MEASURE	TYPE OF SENSOR	GOAL	FREQUENCY OF MEASUREMENT	SAMPLE SIZE	CRITERIA FOR DECISION MAKING	RESPONSIBILITY FOR DECISION MAKING	
Wave solder conditions Solder temperature	Degree F (°F)	Thermo-couple	505°F	Continuous	N/A	510°F reduce heat 500°F increase heat	Operator	…
Conveyor speed	Feet per minute (ft/min)	Ft/min	4.5 ft/min	1/hour	N/A	5 ft/min reduce speed 4 ft/min increase speed	Operator	…
Alloy purity	% Total contaminates	Lab chemical analysis	1.5% max	1/month	15 grams	At 1.5%, drain bath, replace solder	Process Engineer	…
	…	…	…	…	…	…	… …	

FIGURE 4.9 Spreadsheet for "Who does what?" ("Making Quality Happen," Juran Institute, Inc., senior executive workshop, p. F-8, Wilton, CT.)

In this spreadsheet, the essential decisions and actions are listed in the left-hand column. The remaining columns are headed up by the names of the job categories associated with the control station. Then, through discussion among the cognizant personnel, agreement is reached on who is to do what.

The spreadsheet (Figure 4.9) is a proven way to find answers to the long-standing, but vague, question, "Who is responsible for quality?" This question has never been answered because it is inherently unanswerable. However, if the question is restated in terms of decisions and actions, the way is open to agree on the answers. This clears up the vagueness.

PROCESS CONFORMANCE

Does the process conform to its quality goals? The umpire answers this question by interpreting the observed difference between process performance and process goals. When current performance does differ from the quality goals, the question arises: What is the cause of this difference?

SPECIAL AND COMMON CAUSES OF VARIATION

Observed differences usually originate in one of two ways: 1) the observed change is caused by the behavior of a major variable in the process (or by the entry of a new major variable), or 2) the observed change is caused by the interplay of multiple minor variables in the process.

Shewhart called 1 and 2 "assignable" and "nonassignable" causes of variation, respectively (Shewhart 1931). Deming later coined the terms "special" and "common" causes of variation (Deming 1986). In what follows, we will use Deming's terminology.

"Special" causes are typically sporadic, and often have their origin in single variables. For such cases, it is comparatively easy to conduct a diagnosis and provide remedies. "Common" causes are typically chronic and usually have their origin in the interplay among multiple minor variables. As a result, it is difficult to diagnose them and to provide remedies. This contrast makes clear the importance of distinguishing special causes from common causes when interpreting differences. The need for making such distinctions is widespread. Special causes are the subject of quality control; common causes are the subject of quality improvement.

THE SHEWHART CONTROL CHART

It is most desirable to provide umpires with tools that can help to distinguish between special causes and common causes. An elegant tool for this purpose is the Shewhart control chart (or just control chart), shown in Figure 4.10.

In Figure 4.10, the horizontal scale is time, and the vertical scale is quality performance. The plotted points show quality performance as time progresses.

The chart also exhibits three horizontal lines. The middle line is the average of past performance and is, therefore, the expected level of performance. The other two lines are statistical "limit lines."

They are intended to separate special causes from common causes, based on some chosen level of odds, such as 20 to 1.

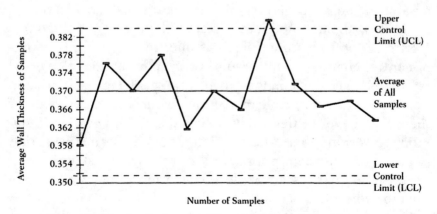

FIGURE 4.10   The Shewhart control chart. ("Quality Control," Leadership for the Quality Century, Juran Institute, Inc., senior executive workshop, p. 4, Wilton, CT)

POINTS WITHIN CONTROL LIMITS

Point A on the chart differs from the historical average. However, since point A is within the limit lines, this difference could be due to common causes (at odds of less than 20 to 1). Hence, we assume that there is no special cause.

In the absence of special causes, the prevailing assumptions include:

- Only common causes are present
- The process is in a state of "statistical control"
- The process is doing the best it can
- The variations must be endured

No action need be taken. Taking action may make matters worse (a phenomenon known as "hunting" or "tampering," or "over-adjusting."

POINTS OUTSIDE OF CONTROL LIMITS

Point B also differs from the historical average, but is outside of the limit lines. Now the odds are heavily against this being due to common causes—over 20 to 1. Hence, we assume that point B is the result of special causes. Traditionally, such "out-of-control" points became nominations for corrective action.

Ideally, all such nominations should stimulate prompt corrective action to restore the status quo. In practice, many out-of-control changes do not result in corrective action. The usual reason is that the changes involving special causes are too numerous—the available personnel cannot deal with all of them. Hence, priorities are established based on economic significance or on other criteria of importance. Corrective action is taken for the high-priority cases; the rest must wait their turn. Some changes at low levels of priority may wait a long time for corrective action.

A further reason for failure to take corrective action is a lingering confusion between statistical control limits and quality tolerances. It is easy to be carried away by the elegance and sensitivity of the control chart. This happened on a large scale during the 1940s and 1950s.

In virtually all such cases, the charts were maintained by the quality departments, but ignored by the operating personnel. Experience with such excesses has led managers and planners to be wary of employing control charts just because they are sensitive detectors of change. Instead, the charts should be justified based on value added. Such justifications include:

- Customer needs are directly involved
- There is risk to human safety or the environment
- Substantial economics are at stake
- The added precision is needed for control

Self-control; controllability.   Employees are in a state of self-control when they have been provided—by management—with all the essentials for doing good work. These essentials include:

- Means of knowing exactly what is expected

  The product or service standard (what is acceptable and unacceptable work)

  The process standard (how to set up and run the job, fill out the form, etc.)

  Who does what and who decides what
- Means of knowing how well one is performing to standard

  Timely feedback on product results (rejects, rework, re-do, scrap, etc.)

  Timely feedback on how the process is running, particularly inputs
- Means to adjust/regulate the process and their performance in the event that performance does not conform to standards

  A process capable of meeting the standards

  A process with features that permit adjustment to bring it into conformance

"Proper" materials, equipment, tools, maintenance, procedures, etc.

Authority to adjust the process.

These criteria for self-control are applicable to processes in all functions and all levels, from general manager to nonsupervisory worker.

It is all too easy for managers to conclude that the above criteria have been met. In practice, there are many details to be worked out before the criteria can be met. The nature of these details is evident from checklists, which have been prepared for specific processes in order to ensure meeting the criteria for self-control. Examples of these checklists include those designed for product designers, production workers, and administrative and support personnel.

If all of the elements of self-control have been met at the worker level, any resulting product nonconformances are said to be worker-controllable. If any of the elements, or even subelements, of self-control have not been met, then management's planning has been incomplete—the planning has not fully provided the means for carrying out the activities within the feedback loop. The nonconforming products resulting from such deficient planning are then said to be management-controllable. In such cases, it is risky for managers to hold the workers responsible for quality, because the workers cannot overcome flaws in the plan.

Responsibility for results should, of course, be keyed to controllability. However, in the past, many managers were not aware of the extent of controllability as it prevailed at the worker level. Studies conducted by Juran during the 1930s and 1940s showed that at the worker level, the proportion of management-controllable to worker-controllable nonconformances was of the order of 80 to 20. These findings were confirmed by other studies during the 1950s and 1960s. That ratio of 80 to 20 helps to explain the failure of so many efforts to solve the companies' quality problems solely by motivating the work force.

EFFECT ON THE PROCESS CONFORMANCE DECISION

Ideally, the decision of whether the process conforms to process quality goals should be made by the work force. There is no shorter feedback loop. For many processes, this is the actual arrangement. In other cases, the process conformance decision is assigned to nonoperating personnel-independent checkers or inspectors. The reasons include:

- The worker is not in a state of self-control
- The process is critical to human safety or to the environment
- Quality does not have top priority
- There is lack of mutual trust between the managers and the work force

PRODUCT CONFORMANCE: FITNESS FOR USE

There are two levels of product features, and they serve different purposes. One of these levels serves such purposes as:

- Meeting customer needs
- Protecting human safety
- Protecting the environment

Features are said to possess "fitness for use" if they are able to serve the above purposes. The second level of product features serves purposes such as:

- Providing working criteria to those who lack knowledge of fitness for use
- Creating an atmosphere of law and order
- Protecting innocents from unwarranted blame

Such product features are typically contained in internal specifications, procedures, standards, etc. Product features that are able to serve the second list of purposes are said to possess conformance to specifications, etc. We will use the short-er label "conformance."

The presence of two levels of product features results in two levels of decision making: Is the product in conformance? Is the product fit for use? Figure 4.11 shows the interrelation of these decisions to the flow diagram.

THE PRODUCT CONFORMANCE DECISION

Under prevailing policies, products that conform to specifica-tion are sent on to the next destination or customer. The assumption is that products that conform to specification are also fit for use. This assumption is valid in the great majority of cases.

The combination of large numbers of product features, when multiplied by large volumes of product, creates huge numbers of product conformance decisions to be made. Ideally, these decisions should be delegated to the lowest levels of organization—to the automated devices and the operating work force. Delegation of this decision to the work force creates what is called "self-inspection."

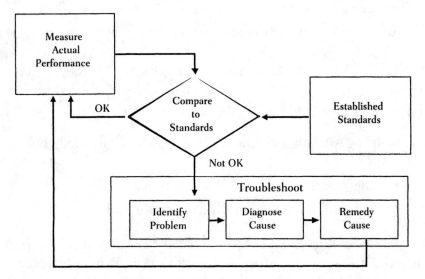

FIGURE 4.11   Flow diagram of decisions on conformance and fitness for use.

SELF-INSPECTION

Self-inspection is a state in which decisions on the product are delegated to the work force. The delegated decisions consist mainly of: Does product quality conform to the quality goals? What disposition is to be made of the product?

Note that self-inspection is very different from *self-control*, which involves decisions on the process.

The merits of self-inspection are considerable:

- The feedback loop is short; the feedback often goes directly to the actuator—the energizer for corrective action.
- Self-inspection enlarges the job of the work force—it confers a greater sense of job ownership.
- Self-inspection removes the police atmosphere created by use of inspectors, checkers, etc.

However, to make use of self-inspection requires meeting several essential criteria:

- *Quality is number one.* Quality must have undoubted top priority.
- *Mutual confidence.* The managers must have enough trust in the work force to be willing to make the delegation, and the work force must have enough confidence in the managers to be willing to accept the responsibility.
- *Self-control.* The conditions for self-control should be in place so that the work force has all the means necessary to do good work.
- *Training.* The workers should be trained to make the product conformance decisions.
- *Certification.* The recent trend is to include a certification procedure. Workers who are candidates for self-inspection undergo examinations to ensure that they are qualified to make good decisions. The successful candidates are certified and may be subject to audit of decisions thereafter.

In many organizations, these criteria are not fully met, especially the criterion of priority. If some parameter other than quality has top priority, there is a real risk that evaluation of product conformance will be biased. This problem happens frequently when personal performance goals are in conflict with overall quality goals. For example, a chemical company found that it was rewarding sales personnel on revenue targets without regard to product availability or even profitability. The sales people were making all their goals, but the company was struggling.

THE FITNESS FOR USE DECISION

The great majority of products do conform to specifications. For the nonconforming products, there arises a new question: Is the nonconforming product nevertheless fit for use?

A complete basis for making this decision requires answers to questions such as:

- Who will be the user(s)?
- How will this product be used?
- Are there risks to structural integrity, human safety, or to the environment?
- What is the urgency for delivery?
- How do the alternatives affect the producer's and the user's economics?

To answer such questions can involve considerable effort. The effort can be minimized through procedural guidelines. The methods in use include:

- *Treat all nonconforming product as unfit for use.* This approach is widely used for products that can pose risks to human safety or the environment—products such as pharmaceuticals or nuclear energy.
- *Create a mechanism for decision making.* An example is the Material Review Board so widely used in the defense indus-

try. This device is practical for matters of importance, but is rather elaborate for the more numerous cases in which little is at stake.

• *Create a system of multiple delegation.* Under such a system, the "vital few" decisions are reserved for a formal decision-making body such as a Material Review Board. The rest are delegated to other people.

Table 4.3 is an example of a table of delegation used by a specific company (via personal communication to one of the authors).

TABLE 4.3   Multiple Delegations of Decision Making on Fitness for Use*

EFFECT OF NONCONFORMANCE IS ON	AMOUNT OF PRODUCT OR MONEY AT STAKE	
	SMALL	LARGE
Internal economics only	Department head directly involved, quality engineer	Plant managers involved, quality manager
Economic relations with supplier	Supplier, purchasing agent, quality engineer	Supplier, manager
Economic relations with client	Client, salesperson, quality engineer	Client: for marketing, manufacturing, technical, quality
Field performance of the product	Product designer, salesperson, quality engineer	Client: managers for technical, manufacturing, marketing quality
Risk of damage to society or of nonconformance to government regulations	Product design manager, compliance officer, lawyer, quality managers	General manager and team of upper managers

*For those industries whose quality mission is really one of conformance to specification (e.g., atomic energy, space), the real decision maker on fitness for use is the client or the government regulator.

DISPOSITION OF UNFIT PRODUCT

Unfit product is disposed of in various ways: scrap, sort, rework, return to supplier, sell at a discount, etc. The internal costs can be estimated to arrive at an economic optimum. However, the effects go beyond money: schedules are disrupted, people are blamed, etc. To minimize the resulting human abrasion, some companies have established rules of conduct such as:

- Choose the alternative that minimizes the total loss to all parties involved. Now there is less to argue about, and it becomes easier to agree on how to share the loss.
- Avoid looking for blame. Instead, treat the loss as an opportunity for quality improvement.
- Use "charge backs" sparingly. Charging the vital few losses to the departments responsible has merit from an accounting viewpoint. However, when applied to the numerous minor losses, this is often uneconomical as well as detrimental to efforts to improve quality.

Failure to use products that meet customer needs is a waste. Sending out products that do not meet customer needs is worse. Personnel who are assigned to make product conformance decisions should be provided with clear definitions of responsibility, as well as guidelines for decision making. Managers should, as part of their audit, ensure that the processes for making product conformance decisions are appropriate to company needs.

CORRECTIVE ACTION

The final step in closing the feedback loop is to actuate a change that restores conformance with quality goals. This step is popularly known as "troubleshooting" or "firefighting."

Note that the term "corrective action" has been applied loosely to two very different situations, as shown in Figure 4.1. The feedback loop is well designed to eliminate sporadic nonconformance like that "spike" in Figure 4.1; the feedback loop is not well designed to deal with the area of chronic waste shown in the figure. Instead, the need is to employ the quality improvement (breakthrough) process of Chapter 8.

We will use the term "corrective action" in the sense of troubleshooting—eliminating sporadic nonconformance.

Corrective action requires the journeys of diagnosis and remedy. These journeys are simpler than for quality improvement (breakthrough). Sporadic problems are the result of adverse change, so the diagnostic journey aims to discover what has changed. The remedial journey aims to remove the adverse change and restore conformance.

DIAGNOSIS OF SPORADIC CHANGE

During the diagnostic journey, the focus is on "what has changed." Sometimes the causes are not obvious, so the main obstacle to corrective action is diagnosis. The diagnosis makes use of methods and tools such as:

- Autopsies to determine with precision the symptoms exhibited by the product and process
- Comparison of products made before and after the trouble began to see what has changed; also comparison of good and bad products made since the trouble began
- Comparison of process data before and after the problem began to see what process conditions have changed
- Reconstruction of the chronology, which consists of logging on a time scale (of hours, days, etc.): 1) the events that took place in the process before and after the sporadic change, that is, rotation of shifts, new employees on the job, maintenance actions, etc., and 2) the time-related product information, that is, date codes, cycle time for processing, waiting time, move dates, etc.

Analysis of the resulting data usually sheds a good deal of light on the validity of the various theories of causes. Certain theories are denied. Other theories survive to be tested further.

Operating personnel who lack the training needed to conduct such diagnoses may be forced to shut down the process and request assistance from specialists, the maintenance department, etc. They may also run the process "as is" in order to meet schedules, and thereby risk failure to meet the quality goals.

CORRECTIVE ACTION—REMEDY

Once the cause(s) of the sporadic change is known, the worst is over. Most remedies consist of going back to what was done before. This is a return to the familiar, not a journey into the unknown (as is the case with chronic problems). The local personnel are usually able to take the necessary action to restore the status quo.

Process designs should provide means to adjust the process as required to attain conformance with quality goals. Such adjustments are needed at start-up and during running of the process. This aspect of design for process control ideally should meet the following criteria:

- There should be a known relationship between the process variables and the product results.
- Means should be provided for ready adjustment of the process settings for the key process variables.
- A predictable relationship should exist between the amount of change in the process settings and the amount of effect on the product features.

CONTROL THROUGH THE REWARD SYSTEM

An important influence on quality control is the extent to which the reward system (merit rating, etc.) emphasizes quality in relation to other parameters.

Experience has shown that control systems are subject to "slippage" of all sorts.

- Personnel turnover may result in loss of essential knowledge.
- Entry of unanticipated changes may result in obsolescence.
- Shortcuts and misuse may gradually undermine the system until it is no longer effective.

The major tool for guarding against deterioration of a control system has been the audit. Under the audit concept, a periodic, independent review is established to provide answers to the following questions: Is the control system still adequate for the job? Is the system being followed?

The answers are obviously useful to the operating managers. However, that is not the only purpose of the audit. A further purpose is to provide those answers to people who, though not directly involved in operations, nevertheless have a need to know. If quality is to have top priority, those who have a need to know include the upper managers.

It follows that one of the responsibilities of managers is to mandate establishment of a periodic audit of the quality control system. Typically, this is the province of the quality assurance function. It is also variously performed by operations managers, and even upper managers. (Please see Chapter 8, "Breakthroughs in Leadership.")

HIGH POINTS OF "THE CONTROL PROCESSES"

- The quality control process is a universal managerial process for conducting operations so as to provide stability—to prevent adverse change and to "maintain the status quo."
- Self-control is of special interest because it provides people all of the resources necessary to do successful work.
- Quality control takes place by use of the feedback loop.
- Each feature of the product or process becomes a control subject-center around which the feedback loop is built.
- As much as possible, human control should be done by the work force—the office clerical force, factory workers, salespersons, etc.
- The flow diagram is widely used during the planning of quality controls.
- The weakest link in facilities control has been adherence to a schedule.

- To ensure strict adherence to a schedule requires an independent audit.
- Knowing which process variable is dominant helps planners during allocation of resources and priorities.
- The work of the planners is usually summarized on a control spreadsheet.
- This spreadsheet is a major planning tool.
- The design for process control should provide the tools needed to help the operating forces distinguish between real changes and false alarms.
- It is most desirable to provide umpires with tools that can help to distinguish between special causes and common causes.
- An elegant tool for this purpose is the Shewhart control chart (or just control chart).
- The criteria for self-control are applicable to processes in all functions, and all levels, from general manager to nonsupervisory worker.
- Responsibility for results should be keyed to controllability.
- If workers are not provided with all the elements of self-control by management, resulting errors are said to be management-controllable, *not* worker-controllable.
- If workers are provided with all the elements of self-control by management, resulting errors are said to be worker-controllable.
- Ideally, the decision of whether the process conforms to process quality goals should be made by the work force.
- There is no feedback loop for controlling processes shorter than self-control.
- To make use of self-inspection to determine product conformance requires meeting several essential criteria: quality is number one; mutual confidence, self-control, training, and certification are the others.
- Personnel who are assigned to make product conformance decisions should be provided with clear definitions of responsibility, as well as guidelines for decision making.

- The proper sequence in managing is first to establish goals, and then to plan how to meet those goals, including the choice of the appropriate tools.

- The planning for quality control should provide an information network that can serve all decision makers.

THE NATURE OF BREAKTHROUGH

CREATION OF BENEFICIAL CHANGE

Breakthrough is a generic umbrella label, referring to the creation of major sustainable beneficial change. Specific types of breakthrough are treated throughout this book. "Breakthrough Improvement" was briefly described in Chapter 2, "The Juran Trilogy," and is thoroughly discussed in Chapter 8, "Breakthroughs in Current Performance." The descriptive label "breakthrough" is employed because creating major sustainable change in an organization can be frustratingly difficult and time-consuming. All of the numerous forces in the organization that resist change must be penetrated and overcome.

PERFORMANCE BREAKTHROUGH

Performance breakthrough refers to Juran Institute's comprehensive approach to achieving and sustaining major organization-wide beneficial change. It consists of the aggregate results of a number of separate, unique prerequisite different types of breakthroughs in various functions and levels in the organization. Breakthrough improvement can produce sudden explosive bursts of localized beneficial change (as from the results of a specific improvement project, e.g., a Six Sigma, DMAIC improvement project). Performance breakthrough, however, may take months or years to accomplish because it is the cumulative effect of many coordinated and inter-related individual improvement efforts. Taken together, these diligent efforts gradually transform the organization. Because of the environment of perpetual unpredictable change in which organizations operate, performance breakthrough efforts must be ongoing, producing continuous adaptive improvements as new needs for improvements impose themselves upon an organization from outside.

Usually a crisis—or a fear of impending crisis—triggers a performance breakthrough effort within an organization.

Consider the following scenario: Our two biggest competitors have introduced some new products that are better than ours. Consequently, sales of product X and Y are heading steadily down, and taking our market share along. Our new product introduction time is much slower than the competition, making the situation even worse. The new plant can't seem to do anything right. Some machines are down often, and even when operating, produce too many costly defective items. Too many of our invoices are returned because of errors, with resulting postponement of revenue and a harvest of dissatisfied customers, not to mention the hassle and costs of re-do. Accounts receivable have been much too high, and are gradually increasing. We are becoming afraid that the future may offer additional threats we need to ward off or, better, prevent.

Management must take action, or the organization is going to experience pain and suffering or worse.

WHY PERFORMANCE BREAKTHROUGH IS ESSENTIAL TO ORGANIZATIONAL VITALITY

There are several reasons why an organization cannot survive very long without the medicinal renewing effects of continual breakthrough.

REASON ONE: COSTS OF POORLY PERFORMING PROCESSES

One obvious reason is that organizations are plagued by a continuous onslaught of crises precipitated by mysterious sources of chronic high costs of poorly performing processes (COP^3). Total chronic levels of COP^3 have been discovered to average 20 to 40 percent or more of costs of goods sold. The figure varies by type of organization.

It is not unusual for COP^3 to exceed profit. In any case, the average overall level is appalling (because it is substantial *and* avoidable), and the toll it takes can be devastating. COP^3 is a major driver of breakthrough initiatives, not only because it can be so destructive if left unaddressed, but also because savings realized by reducing COP^3 go directly to the bottom line. Furthermore, the savings continue, year after year, so long as the remedial improvements are irreversible, or controls are placed on reversible improvements. A case can be made that more improvement to the bottom line can be made by reducing COP^3 than by increasing sales.

It seems only common sense that the mysterious, chronic causes of COP^3 must be discovered, removed, and prevented from returning. Breakthrough improvement becomes a preferred initial method of attack, because of its ability to uncover and remove specific root causes, and hold the gains. It is designed to do just that. One could describe breakthrough improvement methodology as the application of the scientific method to the solution of performance problems. It closely resembles the medical model of diagnosis and treatment.

REASON TWO: CHRONIC CHANGE

Another reason why breakthrough is required for organizational survival is the state of chronic accelerating change found in the environment in which modern organizations operate. Unrelenting change has become so powerful and so pervasive that no constituent part of an organization finds itself immune from its effects for long.

Because any or all components of an organization can be threatened by changes in the environment, if an organization wishes to survive, it likely will be forced into creating basic changes that are powerful enough to bring about accommodation with new conditions. Performance breakthrough, consisting as it does of several specific types of breakthrough in various organization functions, is a powerful approach that is capable of coming up with countermeasures sufficiently effective to prevail against the inexorable forces of change. An organization may have to redesign itself. It may even be driven to re-examine and perhaps modify its core products, business, or service.

REASON THREE: WITHOUT MAKING CONTINUOUS BREAKTHROUGH, ORGANIZATIONS DIE

An additional reason why breakthrough is essential for organizational survival is found in knowledge derived from scientific research into the behavior of organizations. Managers can learn valuable lessons about how organizations function and how to manage them by examining open systems theory. Among the more important lessons taught by open systems theory is the notion of *negative entropy*. Negative entropy refers to characteristics that human organizations share with biological systems such as the living cell, or the living organism (which is a collection of cells). *Entropy* is the tendency of all living things—and all organizations—to head toward their own extinction. Negative entropy consists of countermeasures that living systems and social systems take in order to stave off their own extinction. Organisms replace aging cells, heal wounds, and fight disease. Organizations build up reserves of energy (backlogs) and constantly replace expended energy by acquiring more energy (sales and raw materials) from their environment. Eventually, living organisms lose the race. So do organizations if they do not continually adapt, heal "wounds" (make breakthrough improvements), and build up reserves of cash and goodwill. Breakthroughs are the means by which organizations stave off their own extinction.

PRACTICAL CONTRIBUTIONS OF OPEN SYSTEMS THEORY TO UNDERSTANDING HOW ORGANIZATIONS WORK, AND WHY BREAKTHROUGH IS ESSENTIAL

Open systems theory provides powerful and compelling explanations of how organizations work. These explanations are based on discoveries made by research on the workplace in diverse organizations.

From the vantage point of a researcher, organizations are social systems, and as such, possess certain characteristics, which are summarized succinctly as follows: Open systems continuously interact with and are dependent on their environment. They import, transform, and export energy. Furthermore, in order to survive, they must import from the environment more energy than they expend in transformation and export. As an example, cash reserves must be set aside to outlast hard times, when the level of energy that is imported (sales) is insufficient to maintain normal operations. The cash reserve is energy temporarily stored in the bank until needed to make up for a temporary deficit in imports (sales and accounts receivable). A for-profit organization hopefully will also retain enough cash to realize some profit.

Beyond accumulating and storing excess energy to remain alive, open systems must also perceive and react in adaptive ways to changes in their environment. Their life depends on appropriate and successful transactions with the environment. In short, they must remain open and adaptive.

Somewhat like living organisms, open systems consist of a number of subsystems, each of which performs a vital specialized function that makes specific, unique, and essential contributions to the life of the whole. A given individual subsystem is devoted to its own specific function such as design, production, management, maintenance, sales, procurement, and adaptability. One cannot carry the biological analogy very far because living organisms separate subsystems with physical boundaries and structures (e.g., cell walls, the nervous system, the digestive system, the circulatory system, etc.). Boundaries and structure of subsystems in human organizations, on the other hand, are not physical; they are repetitive events, activities, and transactions. The repetitive patterns of activities are, in effect, the work tasks, procedures, and processes carried out by organizational functions. Open systems theorists call these patterns of activities roles.

> A role consists of one or more recurrent activities out of a total pattern of activities which, in combination, produce the organizational output.

Roles are maintained and carried out in a repetitive, relatively stable manner by means of mutually understood sets of expectations and feedback loops, shown in Figure 5.1. Open systems theory (and Juran's managerial approach) focuses particularly on the human relationships, structures, and interdependence of roles associated with these activities and transactions. A detailed knowledge of the repetitive transactions between the organization and its environment, and also within the organization itself, is essential to accomplishing performance breakthrough, because these transactions determine the effectiveness and efficiency of performance.

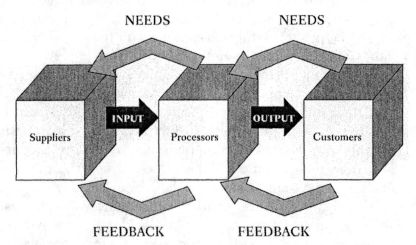

FIGURE 5.1   The triple role (TRIPROL).

THE NATURE OF REPETITIVE HUMAN TRANSACTIONS IN ORGANIZATIONS: HOW WORK GETS DONE AND IS CHANGED

A simplified diagram of the structured repetitive transactions in organizations is found in Figure 5.1.

Figure 5.1 represents a model that applies equally to an organization as a whole, to individual subsystems and organizational functions (departments, work stations, etc.) within the organization, and to individual organizational members performing tasks in any function or level. All these entities perform three more or less contemporaneous roles: they act as customer, processor, and supplier. As a processor, charged with the duty of transforming imported energy, they receive raw materials—goods, information, and/or services—from their suppliers, who may be located inside or outside the organization. The processor's job consists of transforming the received things into a new product of some kind—goods, information, or service. In turn, the processor supplies the product to his/her customers who may be located within or outside the organization.

Each of these roles requires more than merely the exchange of things. Each role is linked by mutually understood expectations (i.e., specifications, work orders, and procedures) and feedback as to how well the expectations are being met (i.e., complaints, quality reports, kudos, and rewards). Note that in the diagram, the processor must communicate (shown by arrows) to the supplier a detailed description of his/her needs and requirements. In addition, the processor provides the supplier with feedback on the extent to which the expectations are being met. This feedback is part of the control loop and helps to assure consistent adequate performance by the supplier. The customer bears the same responsibilities to his/her processors who, in effect, are also suppliers (not of the raw materials, but of the product).

When defects, delays, errors, or excessive costs occur, the causes can be found somewhere in the activities performed by suppliers, processors, and customers, or in the set of transactions between them, or perhaps in gaps in the communication of needs and feedback. Breakthrough efforts must ferret out the precise root causes by deep probing and exploration. If the causes are really elusive, discovering them may require placing the offending repetitive process under a microscope of unprecedented power and precision, as is done in Six Sigma.

Performance breakthrough initiatives also require that all functions and levels be involved, at least to some extent, because each function's performance is inter-related and dependent to some degree on all the other functions. Moreover, a change in the behavior of any one function will have some effect on all the others, even though it may not be apparent at the time.

This inter-relatedness of all functions has practical day-to-day implications for the manager at any level, namely the imperative of employing "systems thinking" when making decisions, particularly decisions to make changes.

SYSTEMS THINKING

Because it is an open system, your organization's life depends on: 1) successful transactions with the organization's external environment, and 2) proper coordination of the organization's various specialized internal functions.

TRANSACTIONS WITH THE EXTERNAL ENVIRONMENT

Organizations import stored energy (in the form of sales, raw materials, equipment, cash, human beings, information etc.). The stored energy is utilized by the production function to produce the goods or services the organization sells, and by all the other functions to support the production function.

Transactions with the external environment are performed by specialized internal functions, which operate outside as well as within organizational boundaries. These functions include marketing and sales, procurement, engineering, finance, human resources, and information technology, all of which support the production function.

The performance of the production function is *dependent* on the performance of all the other functions. The same can be said for the interaction of all functions with each other. They are all inter-related and interdependent. In effect, they all relate with each other as both supplier and customer.

COORDINATION OF THE INTERNAL FUNCTIONS

The proper coordination and performance of the various internal functions, including production, is dependent on the management function (planning, controlling, and improving) and human factors such as leadership, organizational structure, and culture.

To manage an open system (such as your organization), management at all levels must think and act in systems terms: they must consider the impact of any proposed change not only upon the whole organization, but also the impact on the inter-relationships of all the parts. Failure to do so, even when changing seemingly little things, can make some pretty big messes.

Managers need to reason as follows: "If there is to be a change in x, what is required (inputs) from all functions to create this change, and how will x affect each of the other functions, and the total organization as well (ultimate output/ results)?"

Consider this example of the multiple unforeseen impacts on several municipal administrative departments of a well-meant, seemingly simple change made by one department without employing "systems thinking."

The following events occurred in the municipal government of an upscale suburban community located in the metropolitan area of a major U.S. city. Property in this town is expensive and very well maintained. The town rigorously enforces elaborate zoning regulations, not only to enhance the town's appearance—one of its attractions—but also to protect property values.

The burden of enforcing the regulations is shared by several town departments: planning and zoning, building, conservation, engineering, and health. Their procedures, many of which are mandated by state law, also place many responsibilities on property owners and contractors. Planning and zoning (P & Z) plays a major coordinating role.

In order to occupy a new structure or addition, property owners must obtain a certificate of occupancy (CO) from the building department.

Briefly, the process for obtaining a CO looks something like this:

1. Contractor completes construction and calls building (who issued the original building permit), saying the job is done and they want a CO.

2. Building refers the contractor to P & Z, which by state law, must issue a zoning certificate of compliance (ZCC) before a CO can be issued.

3. P & Z, who maintains a file of all these proceedings, refers the applicant to health who must provide a sign-off attesting that the work is in compliance with all health regulations. The sign-off is forwarded to P & Z for the file.

4. If there is a septic system involved, health sends out an inspector who, if everything is OK, issues another sign-off and sends it to P & Z for the file.

5. If wetlands are involved, the applicant must request approval from conservation, who sends out an inspector who eventually forwards a sign-off to P & Z.

6. Now the applicant must ask engineering to perform a grading and drainage inspection, and to forward its sign-off to P & Z.

7. At this point, the property owner must request a surveyor to perform a final "as-built" survey, and send the survey map and documents to P & Z.

8. After all these documents are in the P & Z file, P & Z sends its own inspector to determine that the project conforms to all the zoning regulations. Once everything is in conformance, the inspector creates and files a report.

9. With all the sign-offs, approvals, reports, documents, photographs, etc., in place, P & Z generates a "zoning certificate of compliance" (ZCC), and forwards the ZCC to building.

10. When in possession of the ZCC, building issues a CO.

This is a lot of work to keep track of for the small P & Z staff, you will agree. Under normal conditions, P & Z receives only occasional requests from building for a ZCC, generally on a project-by-project basis, as projects naturally are completed.

Now a minor change is introduced into this long-standing process, with which all departments are familiar, and have staffing and workload plans in place to carry out. During a severe winter, when construction had slowed down, building decided it would be a good time to clean up its old files that contained outstanding building permits for which no ZCC had been received. The projects involved were those which, under normal circumstances, should have been completed during the time that had elapsed since the building permit was issued. Building decided to generate a list, weekly, of several outstanding projects, and refer them to P & Z with a request for a ZCC. (Normally, building would have waited to hear from the contractor that the project was completed before making such a request.)

When P & Z received multiple requests at once, for several ZCCs, this triggered all manner of copying, changes in schedules, correspondence, phone calls, and requests to the contractors, property owners, and other departments for information, and so on.

No one was prepared for this onslaught. Not only was P & Z (who normally is swamped by work it *plans* for) not expecting all this work at once, no one else was, either. Phone calls from anxious and upset contractors and property owners clogged the P & Z phone lines with these questions: "What's happening? What's going on? Have I done something wrong?", etc. Office schedules were thrown off and hastily adjusted. Routine work began to pile up. Everywhere in these departments everything slowed down. Everywhere, that is, except in building, who had executed a well-intentioned plan to make its work more efficient. Here's the kicker: For whatever reasons, building decided to forward not all of the outstanding cases, but only several requests to P & Z at once. Therefore, another list began to arrive every week at P & Z!

What accounts for this mess? A lack of systems thinking:

- Although building made the change for commendable and worthy reasons—to clear out its files and to be efficient—it nevertheless ignored the fact that *other* departments would end up doing the work required to pull it off.
- Building made the change to better meet its own needs, even though it could not by itself execute the change without effort from many others.
- Building failed to coordinate with others who would be effected and would have to do the work.
- Building failed to even inform others who would be affected. The lists just arrived. Surprise!
- Not expecting the surge in work, the other departments could not anticipate and plan to perform the additional work.
- In the absence of joint planning, more effective alternative changes were not explored.
- Citizens and employees of the town were needlessly hassled and stressed.

The preceding is an example of major impact from just one minor change made "the old-fashioned way" of utilizing non-systems thinking, of thinking primarily about the tasks to be done only in my function, rather than the activities required to do the whole job. Contemplate, if you will, the aggregate effects of dozens, hundreds, even thousands of such events as they occur in your organization.

Lessons for Those Seeking to Create Breakthrough

- Problems that appear in one work area or step in a process often have their origin upstream from that work station or step in the process. People in a given work area can't necessarily solve problems in their own work area by themselves.
- Effective and efficient breakthrough can only be created with the active participation, not only of those who are the source

of a problem, but also those affected by the problem and those who will be the source of the changes to provide a remedy to the problem (usually those who are the source of the problem, and perhaps others).

· Breakthroughs attempted in isolation from the whole organization and without systems thinking can easily create more problems than existed at the start of the breakthrough attempt.

Attempts to bring about substantial organizational change such as performance breakthrough require not only changing the behavior of individuals (as might be attempted by training), but also of redefining their roles in the social system. This requires, among other things, changing the expectations that customers have for their processors, and those that processors have for their suppliers. In other words, breakthrough requires actually modifying parts of the organization's very social structure; that is, the set of roles that produce consistent coordinated behavior in support of specific organizational goals. Modifications will likely also be made to other elements that define roles such as job descriptions, work procedures, control plans, other elements of the quality system, training, etc.

To achieve performance breakthrough, then, it is not sufficient simply to train some Six Sigma black belts and do a few projects. Although this will probably result in some improvements, it is unlikely to produce performance breakthrough. The authors believe too many organizations are settling for simple improvements when they should be striving for performance breakthrough.

As we have seen, performance breakthrough consists of making and sustaining major change. It is noteworthy that coming up with a bright idea for a change does not, by itself, make change actually happen. People must actually change what they do, and perhaps how they do it. Beneficial change is often resisted, sometimes by the very persons who could benefit most from it, especially if they have been successful in doing things the current way. Managing change can be a perplexing, challenging undertaking. Accordingly, those trying to implement change should acquire know-how in how to do it. Change is described in Chapter 9, "Breakthroughs in Culture."

BASIC TYPES OF ORGANIZATIONAL SUBSYSTEMS: THE "BASES THAT MUST BE COVERED" TO ACHIEVE PERFORMANCE BREAKTHROUGH

An examination of the various inter-related organizational subsystems will show the clear implications for managers attempting breakthrough.

Organizations (social systems) are composed of the following subsystems:

- A *production subsystem*. Seeks technical proficiency in producing the life-sustaining organizational output of goods, services, or information. The other subsystems support this one. **CAUTION:** Although technical proficiency of the production subsystem may be superb, it does not follow that production and performance will be proportionally flawless. Forces at work by other subsystems may stifle enthusiasm, initiative, cooperation, etc. For example, management practices may create resentment and resistance. If management were to enforce rules (one of their functions) in an arbitrary, inconsistent, or coarse fashion, it may reap sullen withdrawal or resistance instead of cooperation. An even more counterproductive reaction could be expected if the rules themselves are viewed as unreasonable or "stupid."

- *Managerial and maintenance (authority) subsystems*. Seek to provide coordination, stability, and predictability. The managerial subsystem coordinates and controls the other subsystems. It also seeks long-term survival through better use of resources and developing increased capabilities of people and technology. And it resolves conflicts. The maintenance subsystem's mission is to maintain a steady, stable state of affairs.

- A *set of boundary-supportive subsystems such as sales, procurement, disposal, and institutional relations*. Engage in transactions with the environment beyond the organization boundaries to achieve a measure of external control sufficient to support continuing production.

• A *boundary-adaptive subsystem.* Acquires knowledge of changes in the environment and converts the knowledge into recommendations to the managerial subsystem for it to take responsive internal action and make appropriate internal adaptive changes.

> **NOTE:** If not suitably coordinated and controlled, each subsystem tends to take on system characteristics, exhibiting an independent life of its own, complete with its own values, status system, goals, standards, rules, reward structures, etc., which take priority over organization-wide priorities. Each subsystem has a tendency to seek to maximize its own performance, sometimes with little apparent awareness or regard for its own supportive organizational mission, or bad effects on the organization as a whole.

> "Each...subsystem will respond to the same intelligence input in different ways and...each will seek out particular information to meet its needs."
> –Katz & Kahn, p. 228

This situation creates difficulty with transactions and communication across subsystems.

Indeed, it is common for the root causes of performance problems to be found embedded in cross-functional transactions.

THE SIX ESSENTIAL BREAKTHROUGH TYPES WHICH, IN AGGREGATE, HELP MAKE PERFORMANCE BREAKTHROUGH HAPPEN

The following section describes the various types of breakthroughs that, in aggregate, help produce performance breakthrough (see Figure 5.2). Each type of breakthrough addresses a specific organizational subsystem. Each is essential for sup-

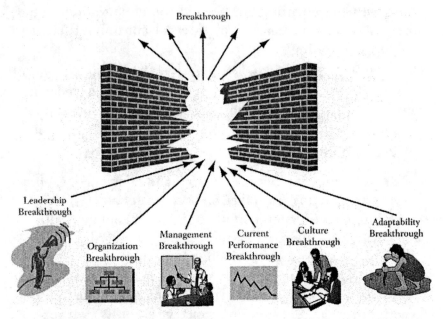

FIGURE 5.2   Performance breakthrough.

porting organizational life; none by itself is sufficient. In effect, they all empower the production subsystem whose mission is to achieve technological proficiency in producing the goods, services, and information for which customers will pay.

Among the different breakthrough types, there is some overlap and duplication of activities and tasks. This overlap is expected because each subsystem is inter-related with all the others, and each is affected by activities in the others.

The authors acknowledge that some issues in each type of breakthrough may have already been addressed by the reader's organization. So much the better. If so, do not start your organization's breakthrough journey from the beginning unless previous efforts were incomplete or poorly executed. Instead, pick up the journey from where your organization finds itself at present. Closing the gaps will likely be part of your organization's next business plan. To close the gaps, appropriate strategic and operational goals and projects to reach those goals will be deployed to all functions and levels. Chapter 11, "Strategic Quality Planning and Deployment," describes this process.

Breakthroughs in leadership.　Breakthrough in leadership is a response to two basic questions: 1) "How do I set performance goals for my organization and activate the people in the organization to reach them?" and 2) "How do I best utilize the power of our people and other resources in my organization and how should I best manage them?"

Issues with leadership are found at *all* levels, not just at the top of an organization. Leadership breakthrough results in an organization characterized by unity of purpose and shared values. Each work group knows what its goals are, and specifically what performance is expected from it. Each individual knows specifically what he/she is to contribute to the overall organizational mission, and how his/her performance will be measured. Little erratic or counterproductive behavior occurs. Should it occur, or conflict arise, guidelines to behavior and decision making are in place to enable relatively quick and smooth resolution of the difficulty.

Top executives and all managers "walk the talk" and lead by example.

Leadership breakthrough does the following:

- Clearly espouses performance goals for an organization
- Clearly articulates and inculcates the mission, values, and norms expected to be reflected in the behavior of an organization's members
- Mobilizes the overall organization to pursue these goals and live by these values

Leadership's basic focus is on change. Leadership breakthrough is described in Chapter 6, "Breakthroughs in Leadership."

The manager's role is one of administering the organization so that high standards are met, proper behavior is rewarded— or enforced, facilities and processes are maintained, and employees are motivated and supported. Performance toward goals is measured and tracked for all functions and levels (i.e., overall organization, function, division, department, work group, and individual). Performance metrics are regularly summarized, published, and reviewed to compare actual performance with goals. Management routinely initiates corrective action to address poor performance, or excessively slow progress toward goals. Actions may include establishing breakthrough improvement projects, providing additional training or support, clearing away resistance, providing needed resources, disciplinary action, etc.

Management does the following:

- Creates and maintains systems and procedures that assure the best, most efficient and effective performance of an organization, in all functions and levels
- Rewards (and enforces, if necessary) appropriate behavior.
- Consistently upholds high standards

Management's basic focus is on stability. Management breakthrough is described in Chapter 6, "Breakthroughs in Leadership."

Breakthroughs in organization.   Breakthrough in organization does the following:

- Designs and puts into place an organization's operational systems (i.e., quality system, orientation of new employees, training, information chains, supply chains, communications, etc.)
- Designs and puts into practice a formal structure that relates each function with all the others and sets forth relative authority levels and reporting lines (e.g., the organization chart)
- Aligns and coordinates the respective interdependent individual functions into a smoothly functioning integrated whole

Organization breakthrough is a response to the basic question: "How do I set up organizational structures and processes to reap the most effective and efficient performance toward goals?"

Trends in this area are clear. More and more work is performed by teams. Job tasks may be described by *team* job descriptions rather than, or in addition to, *individual* job descriptions. Performance evaluation is often related to the accomplishments of one's team instead of or in addition to, one's individual accomplishments.

Management structure consists of cross-functional *processes* that are managed by process owners, as well as vertical *functions* that are managed by functional managers. Where there is both vertical and horizontal responsibility, potential conflicts are resolved by matrix mechanisms that require negotiated agreements by the function manager and the cross-functional (horizontal) process owner.

Unity and consistency in the operation of *both* cross-functional processes and vertical functions is essential to creating breakthroughs, and is essential to continued organizational survival. All members of management teams at all levels simply must be in basic agreement as to goals, methods, priorities, style, etc. This is especially vital when attempting breakthrough improvement projects, because the causes of so many performance problems are cross-functional, and the remedies to these problems must be designed and carried out cross-functionally. Consequently, one sees in a Six Sigma implementation, for example, quality or executive councils, steering committees, champions (who periodically meet as a group), cross-functional project teams, project team leaders, black belts, and master black belts. All of these roles involve dealing with team and teamwork issues.

There is also a steady trend toward fewer authority or administrative levels, and shorter reporting lines. Organization breakthrough is discussed in Chapter 7, "Breakthroughs in Organization."

Breakthroughs in current performance. Breakthrough in current performance (or improvement) does the following:

- Discovers root causes of current chronic problems
- Devises changes to the "guilty" processes that remove or go around the causes

· Installs new controls to prevent the return of the causes.

Breakthrough improvement addresses the question: "How do I reduce or eliminate things that are wrong with my products or processes, and the associated customer dissatisfaction and high costs (waste) that consume my bottom line?"

Breakthrough improvement addresses *quality* problems—failures to meet specific important needs of specific customers, internal and external. (Other types of problems are addressed by other types of breakthrough.)

Quality problems almost always boil down to just a few specific species of things that go wrong:

· Excessive number of defects
· Excessive number of delays
· Excessively long cycle times
· Excessive costs of the resulting reworks, scrap, late deliveries, dealing with dissatisfied customers, replacement of returned goods, loss of customers, loss of goodwill, etc.

Chapter 8, "Breakthroughs in Current Performance," describes how to improve current performance—how to solve quality problems.

Breakthroughs in culture. Breakthrough in culture does the following:

· Creates a set of behavior standards and a social climate that best supports organizational goals
· Inculcates to all functions and levels the values and beliefs that guide organizational behavior and decision making
· Determines organizational cultural patterns such as style (informal versus formal, flexible versus rigid, congenial versus coarse, entrepreneurial/risk-taking versus passivity, rewarding positive feedback versus punishing negative feedback, etc.), extent of internal versus external collaboration, high energy/morale versus low energy/morale, etc.

Breakthrough in culture is a response to the basic question: "How do I create a social climate that encourages organization members to march together eagerly toward the organization's performance goals?"

There are a number of issues to be addressed by breakthrough in culture. Among them are:

- The organization's publicized vision and mission
- Orientation of new employees and training practices
- Reward and recognition policy and practices
- Human resource policies and administration
- Quality policy
- Fanatic commitment to customers and their satisfaction
- Commitment to continuous improvement
- Dress and conduct codes, including ethics
- No "sacred cows" re: people, practices, and core business content
- Organizational citizenship and public relations

Breakthrough in Culture is described in Chapter 9, "Breakthroughs in Culture."

Breakthroughs in adaptability.   Breakthrough in adaptability does the following:

- Creates structures and processes that sense changes or trends in the environment that are potentially promising or threatening to the organization
- Creates structures and processes that evaluate the information from the environment and refer it to the appropriate organizational person or function
- Participates (especially with organization breakthrough) in creating an organizational structure that facilitates rapid adaptive action to exploit the promising trends or avoid the threatening disasters

Adaptability breakthrough is a response to the question: "How do I prepare my organization to respond quickly and effectively to unexpected change?

Adaptability breakthrough is described in Chapter 10, "Breakthroughs in Adaptability."

HIGH POINTS OF "THE NATURE OF BREAKTHROUGH"

- Breakthrough is the purposeful creation of major sustainable beneficial change.
- Performance breakthrough is the aggregate result of achieving a number of prerequisite different types of breakthrough required to create overall (organization-wide) sustainable change in organizational performance.
- Performance breakthrough requires breakthroughs in leadership, management, organization, improvements in current performance, culture, and adaptability.
- Breakthrough is essential for organizational vitality for three reasons (each of which can kill an organization): 1) Costs of poorly performing processes; 2) chronic change; and 3) without making continuous breakthroughs, organizations die.
- To accomplish breakthrough, one must contemporaneously address the three roles each organization, organizational function, and each individual person plays: supplier, processor, and customer, together with the essential associated communications involved in being a supplier, processor, or customer.
- Because your organization, like all organizations, is an open system, your organization's life depends on: 1) successful transactions with your organization's external environment, and 2) proper coordination of the organization's various specialized internal functions.
- In attempting breakthrough, problems that appear in one work area often have their origin upstream in the process. Therefore, people in a given work location suffering from a performance problem can't necessarily solve it by themselves.

- Effective and efficient breakthrough can only be created with the active participation, not only of those who are the source of the problem, but also those affected by the problem and those who will be the sources of remedial changes to the problem (usually those who are the source of the problem and perhaps others).

- Breakthroughs attempted in isolation from the whole organization and without systems thinking can easily create more problems than existed at the start of the breakthrough attempt.

BREAKTHROUGHS IN LEADERSHIP

This chapter addresses two questions:

1. "How do I set performance goals for my organization and activate the people in the organization to reach them?" Issues with leadership are found at all levels, not just at the top of the organization. This chapter reviews characteristics of effective leaders (i.e., those who influence others to follow them), and the essential tasks of leaders: setting goals and strategies to reach the goals. In addition, the chapter points out important differences that distinguish leadership from management.

2. "How do I best utilize the power of our people and other resources in my organization? How do I best manage them? The manager's role is one of administering the organization so that high standards are met, proper behavior is rewarded—or enforced, facilities and processes are maintained, employees are "motivated" and supported, and performance toward goals is monitored and "championed " to remove obstacles. Management holds the organizational structure and operation in place, and continuously plans ahead.

There appear to be two major elements to leadership: 1) You must decide and clearly communicate where you want everyone to go; and 2) You've got to entice them to follow you there.

We use the word "appear" because leadership is a relatively "soft" subject, more of an art than a science. Nevertheless, over the centuries a number of generally agreed-upon characteristics of effective leadership have emerged, and we will discuss these together with some recent findings from research.

First, consider what leadership *isn't*. Leadership isn't management, because management offers incentives for people to behave in the "correct" ways (what the official norms, cultural patterns, and human resources policies say is correct) by providing rewards (cynics might say this is bribery), and threatens people with the possibility of withholding rewards or administering punishments, so they won't behave in "incorrect" ways. Management is mostly concerned with planning and, particularly, stability and control. (In contrast, leadership is mostly concerned with change.)

> **NOTE:** In this book, the words "leader" and "manager" do not necessarily refer to different persons. Indeed, most leaders are managers, and managers should be leaders. The distinctions are matters of intent and activities, not players. Leadership can and should be exercised by managers; leaders also need to manage. If leadership consists of influencing others in a positive manner that attracts others, it follows that those at the top of the managerial pyramid (CEOs and chairmen) can be the most effective leaders because they possess more formal authority than anyone else in an organization. In fact, those at the top usually are the most influential leaders. If dramatic change, such as introducing Six Sigma into an organization is to be undertaken, by far the most effective approach is for the CEO to lead the charge. A Six Sigma launch is helped immensely if other leaders, such as union presidents also lead the charge. The same can be said if senior and middle managers, first-line supervisors, and nonmanagement work crew leaders "follow the leader" and support a Six Sigma launch by word and deed.

Leadership isn't dictatorship, because dictators make people afraid of behaving in "incorrect" ways, and perhaps they occasionally provide public treats (e.g., free gasoline, for example, as has happened in Turkmenistan); freeing prisoners; or staging public spectacles that, together with propaganda, are designed to make people follow the great leader. Dictators don't really get people to want to behave "correctly"(what the dictator says is correct); the people merely become afraid not to.

So what *is* leadership, and leadership breakthrough?

Leadership breakthrough addresses the question: "How do I set performance goals for my organization and activate the people to reach them?"

Leadership breakthrough does the following:

- Clearly espouses performance goals for an organization
- Clearly articulates the mission, vision, values, and norms expected to be reflected, at all levels, in decision making and other behavior
- Mobilizes the overall organization to pursue these goals while believing in and living by these values

Who determines organizational goals? How do they do it? How do they assure that all organization members know and understand not only the goals, but also what each of their individual responsibilities are for meeting the goals? Answers to these questions are discussed in Part I in this chapter.

What do leaders do to get organization members to gradually embrace these values and goals as their own and live by the values while pursuing the goals? Answers to this question are found in Part II in this chapter.

PART I: SETTING GOALS—PLOTTING A COURSE

THE ROLE OF STRATEGIC PLANNING AND DEPLOYMENT

One effective method of establishing organizational goals is strategic planning and deployment. A thorough discussion of strategic planning and deployment is found in Chapter 11, "Strategic Quality Planning and Deployment." A significant benefit resulting from strategic planning and deployment is an organization whose members share common goals.

The upper management team carries out strategic planning annually with the assistance of many others who make special contributions (data, analyses, summary reports, suggestions, proposals, etc.). Deployment is most effective when it involves everyone in the entire organization.

In many organizations, strategic planning and deployment is an integral component of the process that creates the annual budget. Indeed, the strategic plan is the source of significant input into the budget.

Strategic planning.   The first step in strategic planning is to determine the organization's mission. (What business are we in? What services do we provide?) Next, a vision for the desired future state of the organization is formulated and published (e.g., "We will become the supplier of choice, worldwide, of product X or service Y.").

After proclaiming the basic reason for the organization's existence, and the overall general goal the organization seeks to achieve in the future, the upper management team generates a few key strategies the organization is to implement in order to fulfill the mission and realize the vision (e.g., assure ourselves a reliable source of high-quality raw materials; assure ourselves of a stable well-qualified work force at all levels for the foreseeable future; reduce by the end of this year our overall costs of poor quality by 50 percent of last year's annual cost).

Now the process becomes more precise. For each key strategy, a small number of quantified strategic goals (targets) are listed, few enough to be accomplished by the resources and people available. These quantified strategic goals are further divided into goals for this year, goals for the next two years, etc.

Finally, for each quantified strategic goal, a practical number of operational goals is established that describe exactly who is to do exactly what to reach each specific strategic goal. Normally, operational goals are specific projects to be accomplished (such as Six Sigma projects), specific performance targets to be reached by each function or work group, etc.

All of this planning is based on information gathered to inform the upper management team about:

- The organization's current strengths, capacities, weaknesses, and needs
- The competition's strengths, capacities, and weaknesses
- Significant trends that constitute threats or opportunities

Note that organizational goals have been established, not only for the organization as a whole, but also for each of several key strategies, several time periods, several functions, and several levels. Now this plan, considered only preliminary at this point, is deployed.

THE ROLE OF DEPLOYMENT

Deployment is a process of converting goals into specific precise actions, each action designed to realize a specific goal. Deployment occurs in two phases: one phase is *during* the strategic planning process; the other phase is *after* the strategic plan is completed.

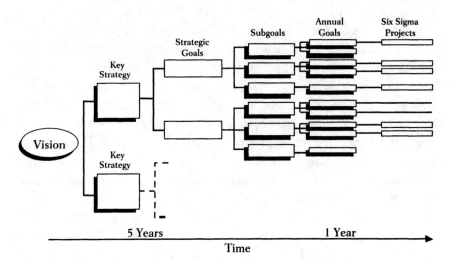

FIGURE 6.1   Juran Institute's model for strategic planning and deployment.

Deployment, phase 1: During the strategic planning process.   After the management team determines key strategies, it circulates these to others in the organization: department heads, functional heads, process owners, etc. They, in turn, may circulate the strategies further out: to supervisors, working foremen, etc. They, in turn, may circulate the strategies to everyone they supervise.

Each party is asked to contribute ideas and suggestions concerning what activities could be undertaken to carry out the strategies, what the specific quantified strategic goals should be, what resources would be required, etc. These responses go back up to the upper management team, who utilizes the responses to promulgate more specific strategic and operational goals.

This exchange of proposed activities to reach goals may take place several times. Some call these iterations and reiterations "catch ball." With each cycle, the various goals are refined, becoming more specific, more practical, and quantified. Finally a set of precise strategic and operational goals emerges, each with owners. In addition, metrics are devised by which to measure performance toward goals and to provide managers at all levels with a score card of progress. Most significantly, these goals have been established with the participation of those who will be responsible for carrying them out.

Deployment, phase 2: After the strategic plan is completed. The published strategic plan is circulated widely. All vice-presidents, managers, department heads, foremen, supervisors, and anyone who supervises others are required to meet with their direct reports. During these meetings, the group reviews the overall plan. They particularly focus on the features of the plan that apply to their work group or to them personally (e.g., their quantified performance or project goals, the specific contributions they are expected to make, the support and assistance they can expect from management, the measurements by which their progress will be reported up the line, etc.).

Emerging from this annual event is an organization united in its commitment to reaching the same goals. All functions and levels have been included. This is highly significant, because leadership is not considered to be something exercised by one person at the top of an organization. It is a function performed at any level and in any function, by anyone who influences others. With a well-deployed strategic plan, specific acts of leadership (attempts to influence others) should be relatively consistent from leader to leader, function to function, and time to time. Decisions made at different levels or in different functions should not conflict with one another very often. So, at least, is the ideal.

In Chapter 9, "Breakthroughs in Culture," the vital importance of consistency in creating an ideal organizational culture is discussed at length.

PART II: GETTING OTHERS TO FOLLOW

PERSONAL CHARACTERISTICS OF LEADERS

Here is where any discussion of leadership becomes fuzzy. There is a wide spectrum of theories, opinion, and schools of thought concerning what are the ideal personal traits a leader should possess. Because of this fuzziness, the discussion of leadership found in these pages is limited principally to *behaviors* exhibited by leaders, rather than personality traits. Managers can at least exert influence on behavior, whereas the ability of managers to influence personality traits is nonexistent.

A few "traits" of effective leaders, nevertheless, emerge repeatedly from experience and the literature. They are probably worth listing. They include:

- Someone who has a background similar to the followers (or at least seems like one of them)
- Someone who is trusted by the followers (the leader tells the truth; the leader makes the followers feel safe, hopeful)
- Someone who is instrumental in producing support or beneficial results for the followers (helps them succeed)

If you reflect on these "traits," it doesn't take long to question whether they are really traits at all, or simply characteris-

tics of behavior attributed to leaders by their satisfied followers. We prefer the latter interpretation.

BEHAVIOR OF EFFECTIVE LEADERS

The reader will recognize that some of the following behaviors
are desirable for managers, as well as leaders, to exhibit. We
quite agree. This is all to the good, since so many leaders are
also managers.

The distinction—to the extent there is one—between managerial and leadership behavior is one of intent: managerial
behavior seeks stability and control; leadership behavior seeks
to produce change.

Leading by example—"walking the talk."   When upper
managers seek to introduce change into an organization, they
should personally participate in the change if they are to truly
lead the effort. The change should visibly affect them. (If it
doesn't affect them, it's: "Do as I say, not as I do," which is
directing, not leading. Directing doesn't enhance the credibility of the upper managers. On the contrary, it lessens their credibility. When that happens, it is an impediment to providing
leadership, because credibility is a prerequisite for leadership.)

For example, if upper managers establish a new Six Sigma
effort that requires many subordinates to spend lots of precious
time in training and in carrying out a project, the upper managers should constitute themselves as a project team, receive
the training, and carry out a project that only people at their
level can tackle. When the rest of the organization sees the
upper management team functioning as a Six Sigma team, the
upper management team's credibility soars, as does their ability to lead.

The results of their project will make a valuable contribution
to reaching strategic goals, and in the process of being trained
and carrying out their project, they will learn enough about the
anatomy of a Six Sigma project to be able to carry out one of
their nondelegable managerial tasks: critiquing progress reports
they receive from all the Six Sigma project teams they have
chartered. Furthermore, they will come to understand typical

barriers to progress that confront all these project teams, such as nonexistent or incapable measurement systems, insufficient or challengeable data, inadequate process controls, and all the frustrating forms of resistance that are aimed at project teams.

Providing everyone the means for success on the job: providing self-control. When managers/leaders do everything they can to provide the means for everyone to attain a state of self-control, this also will greatly enhance their credibility and the level of trust followers will feel toward them. This will happen because when one is in self-control, one has at one's disposal all of the elements necessary to be successful on his/her job. When a leader does this, followers will feel gratitude and respect toward that leader, and will be inclined to follow that leader because: "My leader comes through for me. My leader doesn't just talk; he/she delivers!"

The elements of self-control are discussed in detail in Chapter 6, "Control." A brief review of them follows here, because they can be so instrumental in demonstrating leadership. A person is in a state of self-control, if he/she meets the following criteria:

Criteria for Self-Control

- To know exactly what is expected

  Product standard

  Process standard

  Who does what and who decides what
- To know how he/she is doing compared to the standards

  Timely feedback
- To have the ability to regulate the process

  Capable process

  Necessary tools, equipment, materials, maintenance, time

  Authority to adjust

A person in a state of self-control has at his/her disposal all the means necessary to perform his/her work tasks successfully. Management must provide the means because only management controls the required resources needed to put a person in self-control. Persons who have long been suffering from lack of self-control and its associated inability, through no fault of their own, to perform as well as they would like, are especially grateful to a leader/manager who relieves them from the suffering by making self-control possible. They come to respect and trust such a manager, and tend to become an enthusiastic follower of him/her, mindful of the good things—including enhanced self-confidence and self-esteem—that have flowed from that manager.

Performing periodic management audits.   Conducting periodic audits performed by managers is a superb method of demonstrating commitment to and support for an effort like Six Sigma. Leader/managers, especially the senior executive managers, enhance their credibility and power to lead by personally walking around the organization and talking to the people about what they do, and how they do it.

The management audit has both formal and informal aspects. The formal aspect consists of asking each person being audited to answer certain specific written questions, and to produce data and other evidence of performance that conforms to the formal controls. The informal aspect is simply talking with the folks being audited about what's on their mind, and sharing with them what's on the manager's mind.

The management audit is roughly the equivalent of senior generals visiting the troops in the field. It is a chance for managers to demonstrate their interest in how things are going: what's going well; what needs corrective action. It is a splendid opportunity to listen to what people have to say, and to show respect for them.

If the managers follow up on the suggestions and complaints they hear, that is yet another way to demonstrate that they care enough about "the troops" to provide them with needed support and assistance. It grants to anyone in the organization a direct line of communication with the top, something that makes many people feel important, and motivates them to keep performing at their best. Importantly, the managers' ability to lead is reinforced.

Two examples come to mind. One of the authors was visiting the global headquarters of a huge multinational supplier to the automotive industry. The headquarters building was vast, imposing, surrounded by gorgeous manicured grounds. Its interior lobby was magnificent. At lunch, seated at one of the large round tables in an employee cafeteria, a plain-looking man in rolled-up shirt sleeves, carrying his meal on a tray, asked if he could join us. During lunch, the conversation was lively, down-to-earth, and wide-ranging. The man seemed interested in our opinions on many topics.

It wasn't until the man departed that it was revealed that he was the recently appointed president of the company! He avoided the executive dining room, and took lunch with "the troops." Later, the executive dining room was eliminated, and converted to another employee cafeteria. This man became an effective leader of that immense corporation, effecting many needed changes. He seemed to be just like the rest of the employees.

The same author, while making his first visit to a large new client, was waiting in the beautiful paneled executive suite for the president of this steel company to show up for his appointment. While waiting, an ordinary-looking man dressed in the company's factory uniform and wearing a hard hat strolled into the president's office, looking like someone from Maintenance who came to fix something. Shortly, he emerged, wiping his just-cleaned hands with a towel, introduced himself as the president, and apologized for being late.

He had been delayed by one of his habitual—almost daily—visits to the mills or offices. In this way, he had become familiar with each work station and the people who worked there! He had earned enormous respect from staff in all functions and at all levels, because of these "unpresidential" efforts to understand with his own eyes and ears his employees' daily work life, and their opinions and suggestions. Consequently, he was able to lead that company through a dramatic several-year transformation of its culture and profitability.

Conferring rewards and recognition in public. Leaders can assist their followers to take desired new norms and patterns of behavior upon themselves as their own, if doing so is rewarding to the followers, consistently and over time. The effect of rewards and recognition can be magnified when:

- Rewards and recognition are awarded in public, with fanfare and ceremony
- In the presence of those whose behavior the leader is seeking to influence
- The award is accompanied by an explanation of its connection to a specific desired new behavior toward which the leader is attempting to attract the followers

For example, after launching a Six Sigma initiative, your company decides to have an all-company special assembly to recognize the seven original Six Sigma project teams. Each team makes a presentation of its just-completed project, complete with slides, hand-outs, and exhibits.

After its presentation, each team gets recognition—preferably by the CEO—and perhaps a reward. The person making the award tells the assembled group about the splendid performance of team X. There's more (it gets more specific): how all the team meetings had perfect attendance, how the team carried out all the appropriate steps in the Six Sigma road map, used all the appropriate tools, and produced $150,000 in annual savings, much better satisfied customers, etc.

Not only does the behavior being recognized and praised become reinforced, but all present become better acquainted with the beneficial results that flow from the Six Sigma activities, activities which up until now have been perceived only dimly and with skepticism. Some in the audience begin to be believers. Dr. Juran would say that they begin to get religion after they see the miracles.

NONDELEGABLE MANAGEMENT/LEADER TASKS

When management declares that, henceforth, an organization will be characterized by performance breakthrough, the managers thereby are making personal commitments to perform a number of tasks that only management can perform. In effect, they are signing up to participate in many managerial break-throughs, all of which have been observed to be instrumental in achieving performance breakthrough.

We have already referred to the vital nondelegable leader/manager practices of:

- Setting performance goals
- Deploying goals throughout the organization
- Leading by example
- Providing people the means for self-control
- Conducting management audits
- Conferring public recognition

Several additional practices primarily conducted by man-agers have been identified as vital managerial breakthroughs that are instrumental in producing performance breakthrough.

Nondelegable managerial practices include:

- Creating and serving on an executive council to lead and coordinate the performance breakthrough activities
- Forming policy
- Establishing organizational infrastructure
- Providing resources (especially time)
- Reviewing progress of performance toward goals, including progress of projects
- Removing obstacles, dissolving resistance, providing support and other corrective action if progress is too slow
- Reviewing, and, if necessary, modifying the reward system

Creating and serving on an executive council to lead and coordinate the performance breakthrough activities. An executive council probably exists already in your organization. It consists of the CEO and all direct reports to the CEO. Sometimes other key persons are included: e.g., a union president, a training manager, a Six Sigma master black belt, etc. A performance breakthrough executive council is simply the upper management team. Performance breakthrough activities are simply the way the organization is managed. Performance breakthrough is not something special, not something done on the side if it can be fit into one's "regular" job. Performance breakthrough activities *are* the job, everyone's job. Upper managers in breakthrough performance organizations have consistently been observed to have unique jobs that they have discovered can't be performed by anyone else. It seems the upper managers, by virtue of their station and function in the organization, must personally perform these jobs, which apparently are instrumental in achieving breakthroughs.

The duties of the executive council are essentially those normally associated with managing the enterprise, with a special emphasis on those functions that facilitate continuous change, continuous adaptive improvements. A review of these major duties follows.

Policy formation. This function is strongly inter-related with strategic planning and deployment, as well as breakthroughs in culture and organization. Policies are guides to management action and decision making. They reflect and reinforce organizational values and patterns of behavior. The Executive Council generates product and process quality policies, data quality policies, human resources policies, financial policies, and so on. An effective Executive Council also conducts periodic audits to assure themselves that policies are being followed. The audits may also reveal the need to modify certain policies. All policies should facilitate, not obstruct, the attainment of the organization's performance goals. Obsolete policies that reflect a former time or set of conditions can complicate or even prevent an organization from reaching and maintaining a state of performance breakthrough, and should be changed.

The emphasis of policies should be on the welfare of the organization's customers and clients, as well as its employees. In short, policies should be outward-looking as well as inward-looking, with priority always to outward.

Establishing organizational infrastructure. Chapter 7, "Breakthroughs in Organization," describes this function thoroughly. A pronounced trend is the emergence of teams in the workplace: work-group teams, self-directed work teams, quality improvement teams, product design teams, etc.

The synergy created from a group of individuals working as a team typically produces superior results compared to the same bunch of individuals working as individuals. More information is available to a team than to any individual: a variety of perspectives, a variety of skills, differing perceptions and differing life experiences. Some team members are highly analytical, some are "poets" (highly creative), some focus on relationships, some focus on tasks, and so on.

Managers 1) create; 2) coordinate; 3) evaluate; and 4) critique the performance of teams. In effect, they manage teams, even if from afar in some cases. Because there is considerable know-how required of team members to be an effective team member, managers provide training to teams in how to function as teams. Team leaders receive training in how to lead teams, how to conduct meetings, how to deal with difficult behavior, how to keep a team on track, etc.

There is also considerable know-how involved in managing teams in the aggregate, as does the Executive Council. Consequently, the Executive Council is well advised to provide itself with the same training received by the teams and team leaders. No team is more important than the Executive Council to the life of an organization. Everyone has a vital interest in its harmonious and efficient functioning.

Chapter 3, "The Planning Processes" and Chapter 8, "Breakthroughs in Current Performance" describe additional elements of infrastructure. Those chapters discuss the anatomy (structure, methodology) and functioning of the planning process, the quality improvement process, and the Six Sigma process.

Providing resources. Breakthroughs are produced by teams, project by project. These teams are assigned missions by means of charters. Each project is formally chartered, in writing, by the executive council. (Project team charters are described in

Chapter 3, "The Planning Processes" and Chapter 8, "Breakthroughs in Current Performance.") The Executive Council also provides the project teams with the people and other resources the teams need to carry out their missions.

A list of resources that executive councils typically provide to project teams would include:

- *Project team members.* Members are formally assigned team membership, and this membership, with its associated duties and responsibilities, becomes part of each member's regular job for the duration of the project. This assignment is duly noted in the human resources records, and is taken into consideration during performance reviews.
- *Project team leaders.* Project team leaders also receive a formal assignment, in the manner described above for team members.
- *Project team facilitators.* Six Sigma project teams refer to their facilitators as "black belts." Facilitators/black belts train and coach the project team members and leaders. The job of the black belt is typically full-time, reflecting the importance of the job, and the time required to perform it.
- *Facilities and equipment.* Teams require:

  A quiet, comfortable, well-kept place to meet

  Plenty of wall space on which to post working documents

  A projector

  A screen

  Flip charts and pads

  Stationery supplies

  Telephone

  PC hook-up

  Fax machine

  Usually refreshments
- *A budget.* The precise journey that a project team must take is generally unknown at the launch of a breakthrough project.

The exact route depends on the discoveries made as the project moves forward. Purchases may be required for items such as measuring devices, software packages, services from Information Technology, consulting, training materials, and other special assistance.

- *Time to devote to the project*, at meetings and between meetings. For managers, this is a significant staffing issue. Projects simply cannot make progress without time-consuming effort, particularly by team leaders and facilitators/black belts. Fortunately, this management decision is usually a no-brainer because well-chosen breakthrough projects produce sufficient returns to pay—many times over—the costs of project participants. Occasionally, one observes an organization that expects its project team members to participate in a breakthrough project and also carry on all parts of their regular job, with no provision or allowance made for the several hours of time that such participation requires weekly of team members. Generally organizations that demand this of their people: 1) give the breakthrough efforts a bad name, and reduce support for them; 2) give the managers a bad name because they demonstrate callous disregard for their people and appear to exercise bad judgment; and 3) severely reduce the morale and enthusiasm of the victims of this treatment. Most importantly, the progress of the project and the job performance of the team members both suffer, making matters worse instead of better.

- *Assurance that all organization members will cooperate* with the project teams' efforts, and refrain from resisting or otherwise erecting obstacles. This usually takes the form of a widely publicized and uniformly enforced policy that "everyone will cooperate with breakthrough project teams, and grant their reasonable requests for information, advice, suggestions, etc."

- *Revise the reward system*, if necessary, so performance evaluations grant credit for enthusiastic and faithful participation in breakthrough projects.

Another Vital Managerial Practice:
Make Everyone as Successful as Possible on the Job

We have mentioned the profound contributions that self-control makes to being successful on the job. When managers not only provide all organization members with the means of self-control (even small organizations can do this), but also make other contributions that are dedicated solely to making the individual more successful, the returns can be enormous.

HIGH POINTS OF "BREAKTHROUGHS IN LEADERSHIP"

- Leadership, in this book, includes leading and managing.

- Two major elements of leadership are: 1) deciding where you want everyone to go; and 2) getting others to follow you there.

- The words "leader" and "manager" do not necessarily refer to different persons. Most leaders are managers, and managers should be leaders. The distinctions are matters of intent and activities, not players.

- Setting goals and plotting a course to reach them are activities well carried out by strategic planning and deployment (see Chapter 11, "Strategic Planning and Deployment").

- A significant benefit resulting from strategic planning and deployment is an organization, all of whose members share common goals, understand the organization's goals, and know precisely what contributions each individual is expected to contribute to reaching the goals.

- Effective leaders share a few characteristics. These characteristics include: 1) someone who seems just like the people being led; 2) someone who is trusted by the followers; and 3) someone who is instrumental in producing support or benefits to the followers.

- Behavior of effective leaders includes: 1) leading by example; 2) providing everyone the means for success on the job (providing self-control); 3) performing periodic management audits; and 4) conferring rewards and recognition in public.

- Nondelegable managerial tasks include:

    Creating and serving on an executive council to lead and coordinate performance breakthrough activities

    Forming policy

    Establishing organizational infrastructure (teams, etc.)

    Providing resources (especially time)

    Reviewing progress of performance toward goals, including progress of projects

    Removing obstacles, dissolving resistance, and providing support

    Reviewing, and if necessary, modifying the reward system

    Receiving training and serving on a project team

BREAKTHROUGHS
IN ORGANIZATION

"The rate of change in the business world is not going to
slow down anytime soon. If anything, competition in most
industries will probably speed up over the next few decades.
Enterprises everywhere will be presented with even more
terrible hazards and wonderful opportunities, driven by the
globalization of the economy along with related technological and social trends."

(John P. Kotter, *Leading Change*, 1996)

Getting out of the predicament will require organizations to
adopt an open system approach with four key features:
alignment, integration, knowledge management, and sustainability.

Management deals mostly with status quo and leadership
deals mostly with change. Organizations deal mostly with
structures: functional, cross-functional, teams, etc.

Successful organizations in the twenty-first century will
have to become incubators of leadership. Wasting talent will
become increasingly costly in a world of rapid change.
Developing that leadership will in turn demand flatter and
leaner structures, along with a less controlling and more risk-
taking culture.

ORGANIZATIONAL AND WORK DESIGN: LAYOUT OF WORKPLACES, LEVELS/LAYERS, SHIFTS, AND MORE.

MODERN MANAGEMENT AND ORGANIZATIONAL FADS

When we listen to statements like: "Re-engineering is the search for new models of organizing work. Tradition counts for nothing. Re-engineering is a new beginning" (M. Hammer and J. Champy, 1994), or other even more radical statements by Tom Peters, "Eradicate change from your vocabulary. Substitute 'abandonment' or 'revolution' instead."

Should we really throw away everything that has been done and learned up to his point? Has nothing useful about good management been discovered from the experiences of the great corporations that have transformed life in the past century, as F. G. Hilmer, Dean of the Australian School of Management, puts it? Using a thought from Isaac Newton, in 1676, "If I have seen farther, it is by standing on the shoulders of giants." Hilmer questions, "Is there no one on whose shoulders those interested in learning about and improving the practice of management can stand?" Or is it possible that there is more wisdom in Newton's 300-year-old advice than in contemporary calls for a complete revolution? Cannot true progress come from building on the achievements of the past?

The ideas behind these "gurus," advocates of radical changes, are often labeled "modern management" or a "new paradigm," and are generally considered a vast improvement over traditional notions of a hierarchically organized company with defined processes, management structures, and responsibilities. We do not share this way of thinking.

According to Professor Hilmer, there have been five fads about modern management, but we will consider two of them:

1. Flatten the structure. Hierarchy is passé; flat is beautiful. Modern companies are like orchestras—one conductor and hundreds of players—not armies with long chains of command. Most organizations are hampered by too many levels of management between the board and the frontline

employees who actually invent, make, sell, and provide services. Delayering—downsizing the ranks of middle management in particular—will improve communication, lower cost, speed up decision making, and better motivate all staff to contribute. Less management is better management. And fewer managers are the key.

2. The action approach.

3. Techniques for all.

4. The corporate clan. Model the organization to be more like a happy family than a hierarchy. Create a corporate culture that guides and encourages. Burn the rule books and procedure manuals. Operate as a clan in which people implicitly understand what is right and wrong, and what is good and bad. Rely on the culture to bring out the best in everyone. East is good. West is bad.

5. The board of directors as watchdogs.

The influence of the five false trails on management thought and practice is reflected in language. As Table 7.1 shows, contemporary management-speak has picked up ingredients of each of the trails. Many of these language shifts are positive, a response to changes in the forces that shape business such as increased global competition and new information, communication, and production technologies. But in other respects, the changing language represents a pendulum that has swung too far toward simplifying and inevitably trivializing management, replacing ideas and actions based on sound reasoning with fads and dogma.

TABLE 7.1   The Language Shift

	TRADITIONAL	NOW
Structure	Vertical	Horizontal
	Tall	Flat
	Centralized	Decentralized
	Departments	Teams
	Integration	Subcontracting
	Delegate	Empower

TABLE 7.1   The Language Shift (*Continued*)

	TRADITIONAL	NOW
Action	Planned	Responsive
	Predictability	Ambiguity
	Analysis	Intuition
	Method	Speed
	Hours	Nanoseconds
Techniques	Mass	Niche
	Efficiency	Quality
	Synergy	Breakup value
	Returns	Options
	Goal setting	Benchmarking
	Remuneration	Gainsharing
Culture	Organization	Clan
	Division	Family
	Standards	Shared values
	Formal	Informal
	Diversity	Homogeneity
	Control	Guide
Board of Directors	Monitor	Strategist
	Passive	Active
	Supporter	Critic
	Rule maker	Police

Source: Adopted from "Words by Tom Peters," The Tom Peters Group, California, 1996.

What's wrong with these ideas? Aren't they simply putting into words what the best managers and firms are doing? For example, everyone is flattening structures. Look at General Electric, General Motors, IBM, etc. Or look at newer, hot firms like Nike, Microsoft, or Benetton, where structures have always been flat. Clans and culture are advocated as an alternative to traditional management on the basis of success such as McDonald's, Apple, and Nike.

When examined closely, however, the five trails are shaky or dangerous. They are built on germs of good ideas. But the ideas become false trails when taken too far.

It is difficult to avoid being dragged into these false trails (see Table 7.2). Management is a fertile field for fads and quick solutions because the problems are intractable, yet the pressure to be seen "doing something" is intense. A manager using the latest technique supported by an eminent expert or who is following the widely applauded prescriptions of a best-selling book can hardly be criticized, while one who ignores the latest trends risks being judged as old-fashioned and unprofessional. Moreover, because many management problems are complex and persistent, executives often become frustrated.

TABLE 7.2   Positive Ideas within False Trails

	FALSE TRAIL	POSITIVE IDEA
Structure	Avoid formal structures, hierarchies, and accountabilities: be flexible, ad hoc.	Actively use structure and accountability to direct activities and shape behavior.
	Eliminate levels of management relentlessly: flatter is better.	Beware of levels of management that add no value.
Action	"Just do it."	Keep moving, but be aware of the basis of your actions.
	Follow intuition and gut feelings: end "paralysis by analysis."	Respect analysis, data, and reflection as well as intuition.
	Keep experimenting, trying new ideas.	Distinguish experiments from commitments.
Techniques	Techniques provide effective answers.	Stay abreast of techniques, but be highly skeptical.
	One technique suits most situations.	Customize the techniques you select.
	Keep up with the flow of new techniques.	Limit the number of initiatives underway at any one time.
Culture	Developing and sustaining culture is management's most critical task.	Use clan ideas and culture selectively to reinforce priorities and encourage action.
	Ensure that everyone in the organization adopts the same culture.	Encourage diversity by including various subcultures in the organization.
Board of Directors	Make sure the board keeps management honest and in check.	Focus the board on enhancing corporate and top management performance.
	Independence of directors is key.	Competence and integrity of directors is key.

Source: "The Trivialization of Management" by F. G. Hilmer and L. Donaldson. The McKinsey Quarterly, 1996 Number 4.

The power of Six Sigma is based in most of the positive ideas just described: Actively use structure and accountability to direct activities and shape behavior; respect analysis, data, and reflection, as well as intuition; customize the techniques selected; etc. But above all of them, consistency in the purpose is the most powerful one.

Most managers, however, after a couple of years of effort without bright results, tend to lose confidence in the idea they began with, and start looking for another solution. Nor are investors or stockholders likely to remain patient for long periods. The experience of the Japanese and the Americans with total quality management (TQM) shows it. In the 1960s, the Japanese were working on quality (Dr. Deming and Dr. Juran started to help Japanese managers to understand modern quality principles back in the 1950s), while the Americans were into management by objectives. In the 1970s, the Americans adopted T-Groups and participation, while the Japanese continued to strive for quality. In the 1980s, the Americans shifted to strategic planning, mergers, and restructuring. Meanwhile the Japanese stuck with managing quality. Eventually, the Americans discovered TQM and wondered why the Japanese were so far ahead in managing for quality. The 1990s have been the years in which Americans started to excel again in some industries (mainly using Six Sigma), while the Japanese began to plunge because of their financial problems, although they have not given up quality. The good news is that after 10 or more years of change in major firms, most U.S. companies are catching up and several are forging ahead.

QUALITY PROFESSIONALS IN THE ORGANIZATION OF THE FUTURE

Traditionally, management initiatives, like TQM, self-directed work teams, and quality reengineering have been very much focused on cost reductions. As a result, many organizations have reduced their staff and consequently they have denied themselves the capacity to manage their growth.

To grow, the organization of the future should be agile, sensitive to customers, and able to create knowledge. It should provide value to each of its customers (value is defined as the total benefits the customer perceives along the life of the product or service, minus all the costs during its entire life). Future organizations should continuously exchange information and ideas with customers and suppliers to deliver the products and services requested by each customer. To keep competitiveness, organizations will need to reconfigure their products, services, and processes quickly, and integrate knowledge from other organizations.

So for the organization of the future, it will be critical to create an environment based on innovation and continuous learning, one that can take advantage of the external world uncertainty. The employees should, in turn, have theoretical and practical knowledge to allow them to make decisions by themselves and work in different jobs inside the organization.

The organizations of the future that will succeed will provide value, keeping quality as the paradigm, not as a department. There won't be fewer jobs for quality professionals but more, except that they won't be in the quality department. The word "quality" will not be included on their business cards. These jobs will be part of the value chain of the organization. Quality will be a way of life, not a job or a profession. At the end, it will be clear that quality professionals add value, not costs.

STRATEGY (PLANNING) VERSUS ORGANIZATION (DEPLOYMENT/ IMPLEMENTATION)

Alignment between strategies, financial, technical, human resources, infrastructures, and management systems is the way to make sure adaptation to environmental, technical, or business model occurs.

CENTRALIZED/INTEGRATION/ COOPERATION VERSUS DECENTRALIZED/PARTITIONING/ SEPARATION—INTEGRATION

ORGANIZATIONAL STRUCTURES: COUNCILS, COMMITTEES, TEAMS, AND SELF-MANAGED TEAMS

Organizational structure.   The planning and management controls for each function in a company are intended to guide and support managers and associates in all parts of the organization in their definition of meaningful tasks, and in the specification of procedures to monitor their effective completion. From this perspective, it is easy to recognize the central importance of organizational structure. Responsibility has to do with the nature of the tasks entrusted to each individual. The design process is a primary vehicle to identify, in a coordinated manner, the major tasks faced by the enterprise, and to organize those tasks in the most effective way (Hax and Maluf, 1984).

The organizational structure may be defined as "the relatively enduring allocation of work roles and administrative mechanisms that create a pattern of inter-related work activities and allow the organization to conduct, coordinate, and control its work activities" (Jackson and Morgan, 1978).

There are three accepted basic types of organization for managing any function work, and one newer, emerging approach. The most traditional and accepted organization types are functional, process, and matrix. They are important design baselines because these organizational structures have been tested and studied extensively, and their advantages and disadvantages are well known. The newer, emerging organizational designs are network organizations.

Function-based organization.   In a function-based organization, departments are established based on specialized expertise. Responsibility and accountability for process and results are usually distributed piecemeal among departments. Many firms organize around functional departments having a well-defined management hierarchy. This applies both to the major functions (e.g., human resources, finance, operations, marketing, product development) and also to sections within a functional department. Organizing by function has certain advantages—clear responsibilities and efficiency of activities within a function. A function-based organization typically develops and nurtures talent, and fosters expertise and excellence within the functions themselves. Therefore, it offers several long-terms benefits, and so on. But this organizational form also creates "walls" between the departments. These walls—sometimes visible, sometimes invisible—often cause serious communications barriers. However, function-based organizations can result in a slow, bureaucratic decision-making apparatus, as well as the creation of functional business plans and objectives that may be inconsistent with overall strategic business unit plans and objectives. The outcome can be efficient operations *within* each department, but with less-than-optimal results delivered to external (and internal) customers.

Process-based organizations. Many organizations are beginning to experiment with an alternative to the function-based organization in response to today's "make it happen fast" world. Businesses are constantly redrawing their lines, work groups, departments, divisions, even entire companies, trying to enable productivity increases, cycle-time reduction, revenue enhancement, or an increase in customer satisfaction. Increasingly, organizations are being rotated 90° into processed-based organizations.

In a process organization, reporting responsibilities are associated with a process, and accountability is assigned to a process owner. In a process-based organization, each process is provided with the functionally specialized resources necessary. This has the effect of eliminating the barriers associated with the traditional function-based organization, making it easier to

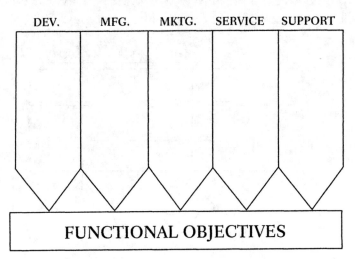

FIGURE 7.1   Functional management system.

Definition by Activity

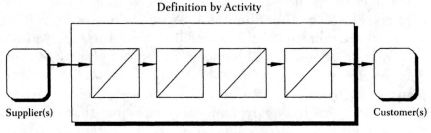

Supplier(s)   Customer(s)

FIGURE 7.2   Business process.

create cross-functional teams to manage the process on an ongoing basis.

Process-based organizations are usually accountable to the business unit or units that receive the benefits of the process under consideration. Therefore, process-based organizations are usually associated with responsiveness, efficiency, and customer focus.

However, over time, pure process-based organizations run the risk of diluting and diminishing the skill level within the various functions. Furthermore, a lack of process standardization can evolve, which can result in inefficiencies and organizational redundancies. Additionally, such organizations frequently require a matrix reporting structure, which can result in some confusion if the various business units have conflicting objectives. The matrix structure is a hybrid combination of the functional and divisional archetypes.

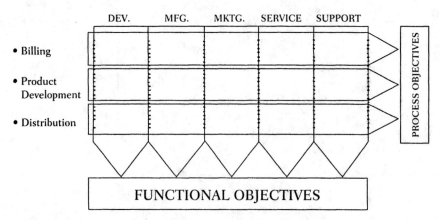

FIGURE 7.3   Functional versus process management system.

Electronic data systems. In 1992, electronic data systems (EDS) put in place a matrix organizational structure radically different from those in most successful U.S. companies (EDS homepage). The matrix structure is a hybrid that has elements of the divisional and network designs. The structure involves 35 to 40 strategic business units (SBUs). The number changes frequently, as it would in a network design, due to business demands.

There are approximately the same number of strategic support centers (SSCs), which support SBU requirements. The organizational relationship between the SBUs and the SSCs form a matrix. Think of the various SSCs as the entries in the columns of the matrix and the SBUs as the entries in the rows of the matrix. Stanford University's business school has included in its core curriculum a case study of this organization design. In this design, key process teams reside organizationally in the SBU, which is the profit center, while commodity production services (e.g., EDS information-processing centers) are outside the profit centers, in the support organizations. The SSCs have the people with the subject matter expertise to perform key functions for the SBUs, and the SBU managers have the budget to pay them for their support services. EDS's approach to human resources is based on customer focus and rewards. These are the key elements of the human resources architecture.

- *Customer focus.* The company keeps its people focused on what is important: quality delivery. The company practices a severe approach to employee alignment: "If you can't change the people, change the people!"
- *Reward system.* Since 1993, EDS has changed its reward system to emphasize team awards. This appears to be comparable to the approach practiced by AT&T's Universal Card Services.

Merging functional excellence with process orientation. What is required, however, is an organization that identifies and captures the benefits of supply chain optimization in a responsive, customer-focused manner, while promoting and nurturing the expertise required to manage and improve continuously the key processes on an ongoing basis.

This organization will likely be a hybrid of the functional and process-based organizations, with the business unit accountable for objectives, priorities, and results, and the functional department accountable for process management and improvement and resource development.

The organization of the future. According to retired Professor F. Gryna, The Center for Quality at the University of Tampa, the organization of the future will be influenced by the interaction of two systems that are present in all organizations: the technical system (equipment, procedures, etc.) and the social system (people, roles, etc.)—thus the name socio-technical systems (STSs).

Much of the research on socio-technical systems has concentrated on designing new ways of organizing work, particularly at the work force level. For example, supervisors are emerging as "coaches"; they teach and empower rather than assign and direct. Operators are becoming "technicians"; they perform a multiskilled job with broad decision making, rather than a narrow job with limited decision making. Team concepts play an important role in these new approaches. Some organizations now report that within a given year, 40 percent of their people participate on a team; some organizations have a goal of 80 percent. Permanent teams (e.g., process team, self-managing team, etc.) are responsible for all output parameters, including quality; ad hoc teams (e.g., a quality project team) are typically responsible for improvement in quality.

The literature on organizational forms in operations and other functions is extensive and increases continuously. For a discussion of research conducted on teams, see Katzenbach and Smith (1993). Mann (1994) explains how managers in process-oriented operations will need to develop skills as coaches, developers, and "boundary managers."

The attributes associated with division managers, functional managers, process managers, and network managers of customer services are summarized in Table 7.3. There is emerging evidence that divisional and functional organizations may not have the flexibility to adapt to rapidly changing marketplace or technological changes.

TABLE 7.3   Attributes of Various Roles in Different Kind of Organizations

	DIVISION MANAGER	FUNCTION MANAGER	PROCESS MANAGER	NETWORK LEADER
Strategic Orientation	Entrepreneurial	Professional	Cross-functional	Dynamic
Focus	Customer	Internal	Customer	Variable
Objectives	Adaptability	Efficiency	Effectiveness	Adaptability, speed
Operational Responsibility	Cross-functional	Narrow, parochial	Broad, pan-organizational	Flexible
Authority	Less than responsibility	Equal to responsibility	Equal to responsibility	Ad hoc, based on leadership
Interdependence	May be high	Usually high	High	Very high
Personal Style	Initiator	Reactor	Active	Proactive
Ambiguity of Task	Moderate	Low	Variable	Can be high

Sources: The first two columns are adapted from the work of Financial Executive Research Foundation, Morristown, NJ. The last two columns represent the work of Edward Fuchs.

DESIGN PRINCIPLES OF WORK AND ORGANIZATION

Work designed for optimum satisfaction of employee, organization, and customer.   Successful organizations are designed to achieve high employee commitment and organizational performance focused on satisfying, and even delighting, the customers. A proper work design allows people to take action regarding their day-to-day responsibilities for customer satisfaction and employee satisfaction.

Customer satisfaction.   Employees must know the customer's needs. They must know whether and how these needs are met, and what improvements can be made to further delight the customer. Employees must have the opportunity to work to ensure that customer needs are satisfied and that performance in this regard is continuously improved. The team must have processes that provide this opportunity. The organization must be designed to ensure that customer needs are known, understood, and communicated, and that there is feedback on the team's performance in meeting the customer's needs. The organizational design should provide the employees the opportunity to work toward continuously improved performance. Improvement must not be expected to happen by accident. An example of such an organization is the Procter & Gamble plant (Buckeye Cellulose Division) located in Foley, FL. The employees at this plant work in self-regulating teams and participate in partnerships with key customers, improving their products and better satisfying customer needs (authors' visit to Foley, CA, 1988).

Employee satisfaction. People naturally want to grow and learn. To enable this to happen, the environment should provide the employees with:

- An understanding of the purpose of the work, strategies for accomplishing it, and the organization's expectations of them
- Adequate pay
- Career growth opportunities
- Adequate authority for their jobs
- Sufficient training and tools
- A feeling of safety in the work environment

The organization must be designed to ensure that employee satisfaction is measured and fed back to the managers. The managers must routinely act on improvement opportunities presented by the employee satisfaction feedback.

Netas is a joint venture in Turkey between the Turkish PTT (the state telephone company) and Northern Telecom Limited of Canada. The Netas philosophy states that a satisfied work force is the key to a satisfied customer (European Foundation for Quality Management, 1995). At Ritz-Carlton, data show that hotels with the highest employee satisfaction levels also have the highest customer satisfaction levels (Davis, 1995). Honeywell and numerous other organizations also recognize the positive correlation between employee satisfaction and customer satisfaction.

In addition to customer satisfaction and employee satisfaction, the organization design must also provide for safe operations, quality and value of products and services, environmental protection, and continuous improvement of processes, products, and people.

Of course, the needs of the business must also be met. The organization must be properly designed to ensure that it meets financial and business objectives, including growth objectives. The design must also enable plants and facilities to operate in a manner that protects the environment, and the health and safety of the employees and the public.

The many business objectives must be defined and communicated so they can be met. A common realistic vision is needed. Again, the organization must be designed to measure performance toward these objectives and provide feedback to the employees.

The way we get this knowledge is to measure the satisfaction of customers, employees, and the organization. Surveys can be very helpful here, as are face-to-face meetings. A balance of measures is needed and usually it is provided through the use of a balanced score card.

The important point of this section is that organizations need to be designed to satisfy customers, employees, society, and investors. Ignoring any one of the stakeholders results in an unbalanced design.

A last word on satisfaction: Miller argues that there is a healthy form of dissatisfaction, which he calls "creative dissatisfaction" (Miller, 1984). It is a healthy thing for employees to be creatively dissatisfied. For example, in a TQM system, where people work together for continuous improvement, failure of the system to perform up to its best every day should cause a healthy level of dissatisfaction. The drive to make things better comes from this creative dissatisfaction with the status quo.

Design a system that promotes high levels of employee involvement at all levels in continuous improvement. Traditional American management was based on Frederick Taylor's teachings of specialization. At the turn of the 20th century, Taylor recommended that the best way to manage manufacturing organizations was to standardize the activity of general workers into simple, repetitive tasks and then closely supervise them (Taylor, 1947). Workers were "doers"; managers were "planners." In the first half of the 20th century, this specialized system resulted in large productivity increases and a very productive economy. As the century wore on, workers became more educated, and machinery and instruments more numerous and complicated. Many organizations realized the need for more interaction among employees. The training and experience of the work force was not being used. Experience in team systems, where employees worked together, began in the latter half of the 20th century, though team systems did not seriously catch on until the mid-1970s, as pressure mounted on many organizations to improve performance. Self-directed teams began to catch on in the mid-1980s (Wellins et al, 1991).

For maximum effectiveness, the work design should require a high level of employee involvement.

Empowerment and commitment. Workers who have been working under a directive command management system, in which the boss gives orders and the worker carries them out, cannot be expected to adapt instantly to a highly participative, high-performance work system. There are too many new skills to learn; too many old habits to overcome. According to reports from numerous organizations that have employed high-performance work systems, such systems must evolve. This evolution is carefully managed, step-by-step, to prepare team members for the many new skills and behaviors required of them. Figure 7.4 presents the steps in the evolutionary process experienced over the period of several years at Eastman Chemical, as described by W. Garwood (Juran's Quality Handbook, 1999). At each new step, the degree of required involvement increases, as does the degree of empowerment conferred on the worker. Many experienced managers have also reported that as involvement and empowerment increase, so, too, does employee commitment to the team, its work, and the long-term goals of the organization.

The directive command is the form most people learn in the military. A command is not to be questioned, but to be followed. It usually results in compliance.

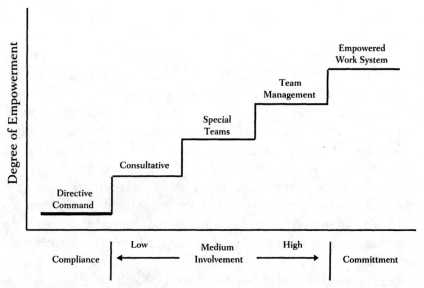

FIGURE 7.4 Relationship of commitment and empowerment.

The first stage of involvement is the consultative environment, in which the manager consults the people involved, asks their opinions, discusses their opinions, then takes unilateral action.

A more advance state of involvement is the appointment of a special team or project team to work on a specific problem, such as improving the cleaning cycle on a reactor. This involvement often produces in team member's pride, commitment, and a sense of ownership.

An example of special quality teams is the blitz team from St. Joseph's Hospital in Paterson, NJ. Teams had been working for about a year as a part of the TQM effort there. Teams were all making substantial progress, but senior management was impatient because TQM was moving too slowly. Recognizing the need for the organization to produce quick results in the fast-paced marketplace, the team developed the "blitz team" method (from the German word for lightning).

The blitz team approach accelerated the standard team problem-solving approach by adding the services of a dedicated facilitator. The facilitator reduced elapsed time in three areas: problem-solving focus, data processing, and group dynamics. Because the facilitator was very experienced in the problem-solving process, the team asked the facilitator to use that experience to provide more guidance and direction than is normally the style on such teams. The result was that the team was more focused on results and took fewer detours than usual. In the interest of speed, the facilitator took responsibility for the processing of data between meetings, thus enabling reduction of the time elapsed between team meetings. Further, the facilitator managed the team dynamics more skillfully than might be expected of an amateur in training within the company.

The team went from first meeting to documented root causes in one week. Some remedies were designed and implemented within the next few weeks. The team achieved the hospital's project objectives by reducing throughput delays for emergency room (ER) patients. ER patients are treated more quickly, and worker frustrations have been reduced (Niedz, 1995).

Special quality teams can focus sharply on specific problems. The success of such a team depends on assigning to the team people capable of implementing solutions quickly. In any TQM system, identifying and assigning the right people is a core part of the system design. Another part of the system design is provision for training teams in problem solving. For many teams, achieving maximum project speed requires a dedicated facilitator.

Team management is the use of teams to manage everyday business, and the continuous improvement of the business. It represents an extension of the authority and responsibility of the natural team.

Natural teams; interlocking team structure: An interlocking team structure creates an environment for sharing ideas and establishing common goals throughout the organization. The concept is that the teams are "natural teams" composed of employees who naturally work together. They are part of a work group, they work within a common work process, they share common goals, and they report to the same supervisor. The team is made up of a supervisor and those reporting directly to the supervisor. The team meets regularly, perhaps weekly or biweekly, to monitor its own performance, and to identify and work on opportunities for continuously improved performance of the team. The supervisor is leader of this team and is, in turn, a member of a team of peer supervisors led by the next-higher-level supervisor. Every employee is, therefore, on at least one team. For examples, at Texas Instruments Europe, most managers belong to at least two quality steering teams, chairing one of them (European Foundation for Quality Management Special Report, 1995).

Implementing team management requires some additional design and training. The design needs to make clear who is on each natural team, when the team will meet, whether participation is required, whether the meetings will require overtime, how the meetings will be organized, and how the team will measure its success.

Each team needs training in group dynamics, team problem solving, statistical thinking, statistical process control (SPC), and team leadership. Leadership is usually rotated every 6 to 12 months. Therefore, some training is needed to prepare people for their assignments. As teams mature, they will need training in providing feedback to each other.

Dr. Juran has predicted that the next improvements in organizations will be widespread introduction of empowered work systems, including teams variously labeled as high-performance teams, self-managing teams, self-directed teams, empowered teams, natural work groups, and self-regulating teams, with self-directed work teams being the most common. A self-regulating team is a permanent team of employees empowered to manage themselves and their daily work. Team responsibilities may include: planning and scheduling or work, resource management, budgeting, quality control, process improvement, performance management, and other activities related to human resources. Thus, these teams perform some activities that have been traditionally performed by supervisors. The supervisor is still involved, but now serves as a coach. Self-regulating teams manage their business.

These are not ad hoc teams. They are permanent teams, frequently organized around a work process. They see themselves as "entrepreneurial business units," providing value-adding services, responsible for the satisfaction of their customers. For existing organizations that make the transition to empowered teams, some work system redesign is usually required.

These teams will take the organization to even higher levels of performance. Members of the empowered team support each other and the organization to achieve organizational and team goals (which, many times, they have helped establish). Peters states that the self-managing team should become the basic organizational building block (Peters, 1987). Empowered work systems are designed so that the teams are empowered (having capability, authority, desire, and understanding of their purpose), and thus positioned to meet the needs of customers, business, and employees.

Empowered work team membership can make 80 to 90 percent of daily decisions regarding the team's business. The team is responsible for its own actions and team results, and thus for the overall product and service provided. Members accept that their job is expanded to include improving the work processes of the team. They also accept more responsibility for problem solving, leadership, and team development as the team matures. Administrative tasks such as scheduling and training are coordinated by the team itself. (Good coaching is required during the maturing process.)

The team receives cross training to make team members multiskilled and provide the flexibility needed to meet changing customer needs with high quality outputs. Through the cross training to provide "multiskills," team members can typically perform two to three different jobs. Multiskilling creates a more flexible organization, therefore, a more efficient one. Associated with this is the ability to respond more readily to changing customer needs and to operate with a reduced number of people, compared to the number required in the absence of multiskilling. A survey of innovative organizations indicated that new "Greenfield" organizational start-ups or new plant designs usually exhibit "50 to 70 percent" better performance (profitability) than a "traditionally" designed organization and facility (Macy, 1995).

Empowered, self-regulating teams monitor their progress and work on redesigning their work processes with regular redesign steps in 12- to 24-month intervals. Teams at the Procter & Gamble plant in Cape Girardeau, MO, conduct regular redesigns every two years to capture lessons learned of things that have worked and did not work in the previous work designs. These teams had been in operation for over 10 years, and have done several redesigns for continual improvement. This continuous improvement keeps what works best and modifies things that do not work as well. Coaching for empowered teams is very important, particularly during the first two to three years of operation.

Empowered teams distinguish themselves from traditional work groups in numerous ways. An example is the lengths to which a team goes to understand what satisfies and dissatisfies the team's customers. In one company, one team sent several team members on visits to learn customer reactions first-hand, directly from the workers who used their product. They reported their findings to the team, using video footage made in customer operations by a team member.

Empowerment principles are applied in the engineering environment, where employees work in cross-functional teams to design a new plant or product in minimum time and with maximum effectiveness. Sometimes these teams include suppliers, customers, and the necessary company functions, as well as contractors. Toyota, Ford, Honda, and General Motors are among the companies that have effectively used such teams to design new automobiles in record times. Design activity within such a team permits rapid reaction throughout the organization to design proposal, including early warning of impacts up- and downstream from the proposed change. A proposed change—a new supplier, a tightened dimensional tolerance, a new paint material—can quickly be evaluated for its effect on other operations. This aspect is often referred to as "simultaneous engineering."

These principles are applicable at the highest levels of the company. In one company, a group of division presidents operates as a self-managed group of peer managers, directing worldwide manufacturing efforts. These managers meet regularly to develop work processes and monitor measures of success. They rotate leadership. Ownership and commitment are evident here, just as in a team of operating technicians.

A summary of the most common types of quality teams is given in Table 7.4.

TABLE 7.4   Summary of Types of Quality Teams

	SIX SIGMA/ QUALITY PROJECT TEAM	EMPLOYEE INVOLVEMENT/ QUALITY CIRCLE	BUSINESS PROCESS QUALITY TEAM	SELF-DIRECTED TEAM
Purpose	Solve cross-functional quality problems	Solve problems within a department	Plan, control, and improve the quality of a key process	Plan, execute, and control work to achieve a defined output
Membership	Combination of managers, professionals, and work force from multiple departments	Primarily work force from one department	Primarily managers and professionals from multiple departments	Primarily work force from one work area
Basis and Size of Membership	Mandatory; 4-8 members	Voluntary; 6-12 members	Mandatory; 4-6 members	Mandatory; all members in the work (6-18)
Continuity	Team disbands after completion of project	Team remains, project after project	Permanent	Permanent
Other Names	Quality improvement team; Six Sigma improvement team	Employee involvement group	Business process management team; process team	Self-supervising team; blitz team; self-directed work teams

Source: Juran's Quality Handbook

Teams.   What is the key for the best teams? It is usually dedicating some time to organize the work that should be accomplished by the team. Time is necessary to organize and make sure you and your team know what you're doing, why you are doing it, how to organize the work, and who is involved. However, the schedules are so short, there is never time to organize the work team.

Many teams start working, believing they know what to do, and so go into action directly. They don't have a minute for finding the support from others, or determining the right goals for the team, or to create and implement a plan that allows them to achieve the proposed goals or to plan how to work together. Nevertheless, what they all have in common is that no one understands in the same way what they are doing, why, how, and with whom. Linking the effort of the team to five critical success factors can solve the problem.

Five critical success factors

1. *Have a shared conscious goal.* Know the goal of the team, why the work is necessary. Be conscious of the consequences if it doesn't work. See what can be achieved and know what can motivate the team.

2. *Set challenging, but specific goals.* What are the key results we commit to get if we achieve the goal? What are the aims and obstacles that we need to identify to avoid deviations from our goal and achieve what we are looking for? When? What kind of hitch should we expect?

3. *Get a common approach so that everyone can participate.* Raise questions like: How will we achieve our objectives? What is our project? Which method will we use? Which key issues should we consider to be successful as a team? Which will be our strategy to get third-party support?

4. *Clear competencies.* How are tasks and activities going to be distributed and performed in the team? What's the role of the project leader, the facilitator, and the team members? Who and how will they make key decisions?

5. *Complementary abilities.* How can the team be established to combine knowledge with experience and skills to make the work efficient? How are we going to use the complementary abilities to support all of us? How are we going to lead with mistakes?

The answers to all these questions provide the framework that supports an effective collaboration to make a team work. Once the plans for the projects to work on are decided by agreement, team members feel much more confident about the commitments of the project goals to support each other and to achieve their obligations.

Preparing a plan helps as a road map for how different people and teams will contribute to the project, as it has all components: involves shareowners, helps achieve the agreements of each phase without deviations from the main goal, and allows to attain the desired results.

The team charter includes the agreements from the answers to key questions: what, why, how, and who. As specific and clear goals are established in the team, you get a shared vision of the project and the work directions for the team. You will be able to establish clear expectations when it comes to assess team progress, as beforehand the main goals for the team have been defined. The charter will help as a road map to avoid getting lost with the work to be done and to determine when it has concluded.

Leadership style.   Members of empowered teams share leadership responsibilities, sometimes willingly and sometimes reluctantly. Decision making is more collaborative, with consensus as the objective. Teams work toward win-win agreements. Teamwork is encouraged. Emphasis is more on problem solution and prevention, rather than on blame. During a visit to Procter & Gamble's plant in Foley, FL, the host employee commented that a few years before he would not have believed he would ever be capable of conducting this tour. His new leadership roles had given him confidence to relate to customers and other outsiders.

Citizenship.   Honesty, fairness, trust, and respect for others are more readily evident. In mature teams, members are concerned about each other's growth in the job (i.e., members reaching their full potential). Members more willingly share their experiences and coach each other, as their goal is focused on the team success, rather than on their personal success. Members more readily recognize and encourage each other's (and the team's) successes.

Reasons for high commitment. As previously stated, empowered team members have the authority, capability, desire, and understanding of the organization's goals. In many organizations, they believe that this makes members feel and behave as owners, and makes them more willing to accept greater responsibility. They also have greater knowledge, which further enhances their motivation and willingness to accept responsibility.

Means of achieving high performance. It has been observed that as employees accept more responsibility, have more motivation, and greater knowledge, they freely participate more toward the interests of the business. They begin to truly act like owners, displaying greater discretionary effort and initiative.

As previously stated, empowered team members have authority, capability, and desire and understand the organization's direction. Consequently, members feel and behave as owners, and are willing to accept greater responsibility. They also have greater knowledge, which further enhances their motivation and willingness to accept responsibility. An empowered organization is contrasted to a traditional organization in Table 7.5.

TABLE 7.5 Traditional versus Empowered Organization

ELEMENT	TRADITIONAL ORGANIZATION	EMPOWERED ORGANIZATION
Guidance	Follow rules/procedures	Actions based on principles
Employee focuses on	Satisfying the supervisor	Satisfying the customer, and achieving the business objectives
Operator flexibility	One skill	Multiple skills
Participation	Limited	High involvement
Empowerment	Follows instructions, asks permission	Takes initiative; a can-do attitude; discretionary effort
Employee viewed as	A pair of hands to do defined tasks	Human resource, with head, heart, hands, and spirit
Leadership for work processes	Managers only	Shared by managers and operators
Management communication style	Paternalistic	Adult to adult

TABLE 7.5   Traditional versus Empowered Organization
(*Continued*)

ELEMENT	TRADITIONAL ORGANIZATION	EMPOWERED ORGANIZATION
Responsibility for continuous improvement	Management	Shared by managers, staff, and operators
Work unit defined by	Function (such as manufacturing or sales)	The work process, which may be cross-functional
Administrative decisions are made by	Management	Shared responsibility of team members and management (if self-directed team, the team may be responsible solely for certain administrative decisions)
Quality control is the responsibility of	Laboratory	Team capability for process control

Source: Eastman Chemical Company

Key features of an empowered, high-performance organization.   An outstanding example of an organization that exemplifies these characteristics is SOL, a Finnish building cleaning company.

Creating an environment.   Liisa Joronen, head of SOL, believes that an organization must give its employees every opportunity to perform at their best. To achieve such a state, she has taken steps to create an environment that gives SOL workers whatever they need to get the job done. This includes:

· Complete freedom to work when, where, and how an employee chooses, so long as customer needs are met. Workers are given cellular phones, voice mail, e-mail, laptop computers, and home computers as needed, resulting in much easier access to people than traditional set-ups allow.

· The absence of organization charts, job titles, and status symbols, including secretaries and company cars (Joronen often rides her bicycle to meetings with external customers). Each office worker has an area of primary responsibility, with the understanding that he/she will also contribute in other areas, as vacations or other conditions warrant.

- An open-book policy on company performance. Each month, employees receive updated information on financial figures, absenteeism, turnover, and the like, right down to individual performance among the staff.

- A clean desk policy, which eliminates territorial office claims. Employees use whatever work space is available. When the work is done, the employee clears the area of his/her materials so that others may use it (Juran Institute, 1996).

Over the past 20 years, enough progress has been made with various empowered organizations that we can now observe some key features of successful efforts. These have come from experiences of various consultants, visits by the authors to other companies, and published books and articles. These key features can help us learn how to design new organizations or redesign old ones to be more effective. The emphasis is on key features, rather than a prescription of how each is to operate in detail. This list is not exhaustive, but is a helpful checklist, useful for a variety of organizations.

Focus on external customers.   The focus is on the external customers, their needs, and the products or services that satisfy those needs.

- The organization has the structure and job designs in place to reduce variation in process and product.
- The organizational layers are few.
- There is a focus on the business and customers.
- Boundaries are set to reduce variances at the source.
- Networks are strong.
- Communications are free-flowing and unobstructed.
- Employees understand who the critical customers are, what their needs are, and how to meet the customer needs with their own actions. Thus, all actions are based on satisfying the customer. The employees (operator, technicians, plant

manager, etc.) understand that they work for the customer rather than for the plant manager.

- Supplier and customer input are used for managing the business.

Guidance is by principles

- There is a common vision, which is shared with and understood by all employees in the organization.
- All actions and decisions are based on a stated philosophy, which refers to the organization's mission, values, and principles.

Dana Corporation provides a good example. In the early 1970s, Dana eliminated company-wide procedures, replacing them with a one-page statement of philosophies and policies. Dana states, "We do not believe in company-wide procedures. If an organization requires procedures, it is the responsibility of the appropriate management to create them" (Dana Corporation, 1994).

There is a relentless pursuit of continuous improvement and innovation.   Robert Galvin, former CEO of Motorola and Chairman of the Executive Committee, has said that the most important quality lesson he learned in his years at Motorola was that "perfection is attainable" (Galvin, 1996). He did much to lead Motorola on the path to perfection, through continuous improvement and innovation.

The Malcolm Baldridge National Quality Award, established by the U.S. Congress in 1987, presented annually by the President of the United States, is itself an example of continuous improvement:

- The organization and its supporting systems encourage all employees to improve products, processes, teams, and themselves
- Continuous learning is part of the job
- Stretch improvement goals are established. Examples include Motorola's Six Sigma and Xerox's 10x improvement goals

- Informed risk taking is encouraged
- A systematic means exists for periodic organization renewal
- Coaching and development systems are in place for all teams and individuals

Management can encourage improvement by giving appropriate recognition, such as by special awards and celebrations for achieving incremental performance goals, for reaching new records, etc.

Shared leadership: there are new roles for both operators and managers.   Supervisors in empowered, high-performing organizations find themselves in new roles, which include coaching and developing teams and individuals, clarifying business expectations and responsibilities, managing the interface between teams and their environment, allocating resources among teams, and ensuring that continuous improvements are occurring. All of this represents a span of responsibility greatly increased beyond that of the traditional supervisor.

Managers make provision for everyone's input to decisions affecting the larger organization, and for participative planning processes at the functional level.

Another perspective on leadership comes from Netas, and may be the harbinger of an emerging principle of management in empowered organizations. Managers at Netas believe that leaders can come from anywhere in the organization, that the organization has in it many leaders, and that a leader creates leaders. To foster leadership, top managers need to be visibly involved. Top managers at Netas contribute to leader development by leading teams, chairing all customer and distributor conferences, and attending presentations of continuous improvement teams (European Foundation for Quality Management Special Report, 1995).

Operators also find themselves in new jobs, such as scheduling work, hosting suppliers, visiting customers, working on cross-functional process teams, and leading the search for root causes.

The team concept emerges.   The emerging team concept includes the following features:

· Work force empowerment
· Control of variances at the source
· Sharing of leadership and responsibilities by supervisor and team
· Agreed-on, well-defined behavioral norms for individuals and team
· Constructive peer feedback for team and individual development
· Performance measures that enable the team to monitor its performance and take actions when needed
· Team responsibility for finished products or services
· Team participation in the process of hiring new team members

Empowered organizations have many support systems in common, which enable them to function well. These include:

· Reward and recognition of desired skills and behaviors
· Business and accounting information to support decision making at the point of action
· Systems to obtain, maintain, and develop qualified personnel
· Leadership to enable teams to achieve their mission

THE ROLE OF MANAGEMENT IN SUPPORTING AN EMPOWERED ORGANIZATION

In empowered organizations, managers create an environment to make people great, rather than control them. Successful managers are said to "champion" employees and make them feel good about their jobs, their company, and themselves. Marvin Runyon, when head of the Nissan plant in Smyrna, TN, stressed that "management's job is to provide an environment in which people can do their work" (Bernstein, 1988).

The role of management includes the following:

- *Create a vision of the business, and share it widely.* The Baldrige Award Category 1 states (and numerous leaders agree) that all employees must have a shared vision of the organization as an empowered, high-performance organization, which satisfies its customers, is efficient and effective, and works toward continuous improvement. Management's job is to create, share, and maintain this vision. Deming reminded us of the need for constancy of purpose in achieving the long-term objectives for the organization.

- *Set the organization objectives and strategies*, and share them.

- *Structure and align the organization* to achieve the strategies. Create a role for everyone in the business.

- *Allocate resources* (including efforts such as research and education).

- *Communicate* business information.

- *Listen for the needs of the organization.* Encouraging open, two-way communication, which is essential for employees to contribute fully toward the organization's objectives, is another management task. At the Dupont plant in Asturias, Spain, managers visit team meetings and encourage team leaders.

- *Create the environment* for sharing ideas and forming common goals throughout the organization.

- *Create* interlocking team structures.

- *Identify and eliminate barriers to teamwork.* Symbols that differentiate employee groups can be powerful barriers to teamwork. Such symbols include dress codes, car parking privileges, special dining rooms, clock cards, and special vacation policies, to name several. At the DuPont plant in Asturias, Spain, much attention has been devoted to eliminating divisive symbols. Every employee has a business card, no matter what the job level is, but no titles or other indications are given to differentiate job level. Office furnishings are identical for each person with an office. Every employee dresses consistently, with no differentiation. The traditional performance rating system can also be a barrier to teamwork. Eastman Chemical has eliminated its traditional perform- ance appraisal system, which had focused on annual ratings for employees, and replaced it with a system focusing on employee growth and development. The new system is designed to foster cooperation rather than competition among employees. Any disparities in benefits from one group to another can create a barrier to teamwork. Such disparities should be identified and removed. Employee suggestion sys- tems that emphasize individual reward for suggestions create conflicting incentives for team members. Self-interest may drive a team member to reserve valuable suggestions for the channel that provides the greatest reward. Such conflicting incentives create insurmountable barriers to team success and must be addressed before launching team efforts.

- *Emphasize and support training.* When there is sufficient staffing, employees can receive needed training without requiring the team to work overtime. Some operations have designed their work system to include an extra crew, to pro- vide work coverage while some personnel fulfill training needs.

- *Pay for skills and knowledge.* The pay system is carefully designed to recognize and encourage multiskilling. As employees learn additional skills and knowledge, they earn more pay. Compensation for team members may comprise up to four elements: base pay for the job; pay for individual per- formance, skill, and knowledge; pay based on the perform- ance of the team; and pay based on the organization's overall performance. In the context of empowered teams, the sim- plest pay system is to pay a base level, and then achieve moti- vation through intrinsic factors of the job design and sharing gains in productivity.

- *Assure job stability.* When improved performance will render certain jobs unnecessary, employees may resist fully con- tributing in continuous improvement efforts. With its transi- tion to empowered, self-regulating teams, Eastman eliminated over 50 first-line supervisory jobs. However, no one was laid off; former supervisors were eventually absorbed into other jobs within the company, often with the help of retraining. Many excess supervisors became team facilitators and coaches. During the transition, some excess supervisors became part of a "redeployment pool," assigned special projects, even for a few years, while waiting for appropriate job opportunities to open up, often through attrition. Most of these redeployed supervisors later admitted that life got better for them as a result of their new opportunities.

REINFORCE POSITIVE BEHAVIORS

A critical role of management is to reward and reinforce behaviors and results that help achieve the overall goals set for the organization. This reinforcement reminds employees that these behaviors and results are truly important. Without this reinforcement, employees may come to believe that their actions and accomplishments are not really important because they resulted in no positive personal consequence. They may then wonder if something else has become more important, which leads to reduced focus on the organization's objectives.

ORGANIZATIONAL SYSTEMS: QUALITY SYSTEMS AND EFFICIENCY MEASUREMENTS

- *Systems.* Way of working, organizing, or doing something in which you follow a fixed plan or set of rules

  Government or administration

  Way in which it is arranged so that all of its parts fit together or work together
- *Efficiency.* Being able to do a task successfully without wasting time, energy, or cost
- *Effectiveness.* Working well and producing the results that were intended, particularly meeting vital needs of vital customers

ORGANIZATIONAL SYSTEMS: IT, TECHNOLOGY, AND MORE

ORGANIZATION AND TECHNOLOGY

Modern armies base their efficiency in the utilization of information technologies and in the management of a flat organization instead of a hierarchical one.

Today, the competitive advantage in wars is based not so much on the destructive capacity of the weapons, but on the ability to manage the human and technical resources. We all know technology has made weapons much more efficient, but the biggest impact has been the application of IT to the management of an armed conflict. In the last war against Iraq, the United States used a technology called Theater Battle Management Core System (TBMCS), which is the core of network-centric warfare and uses broadband and network software in a way in which the information obtained from surveillance elements, vehicles without crew, geospace information, and human intelligence is captured, processed, and distributed in real-time to the military people on the battlefield.

This is the dream of every company: be able to gather, process, and send to their executives and employees, in real-time, all the information regarding the marketplace, the customers, the suppliers, production, stocks, and competitors. (See Chapter 10, "Breakthroughs in Adaptability.")

What are the new problems that emerge from this scenario? Mainly two. First is the reliability of the information and communications systems. In very tough and adverse conditions systems must be transported to the battle scenario with all the security controls in place to avoid hackers or access through the passwords obtained from prisoners. The second, and probably more complicated problem, is the necessary change of mind in the army in migrating from a perfectly hierarchical organization to a flat one, with decentralized decisions based on the information needed in real-time.

These are also the two problems for many companies. The dependency on the information and communication technologies is so high that when a defect happens, there are no alternative processes capable of allowing continuity to operations. Changing the mindsets of all employees—from just executing to taking decisions—requires a recycling program, rather than just a training program.

In any case, network-centric warfare is already here and, as many other military innovations, will have an impact on other types of organizations. The first one would be the confirmation of the importance of integrating, processing, and distributing all information to allow employees to take consistent decisions (empowerment), aligned with the company goals. Companies that can move to this new scenario will have competitive advantages. The second one is that the need for feasible security of the information—on a battlefield, in which each soldier, in his/her tank, plane, or even walking, can have access to a huge amount of data in real-time—will drive a technological development that will also be applicable to all types of organizations.

ORGANIZATIONAL SYSTEMS: KNOWLEDGE MANAGEMENT

ORGANIZATION AND KNOWLEDGE MANAGEMENT

Broken down into its simplest form, the learning process consists of observation-assessment-design-implementation. These can vary along two main dimensions:

- *Conceptual learning.* The process of acquiring a better understanding of cause and effect relationship, leading to "know-why."
- *Operational learning.* The process of obtaining validation of action outcome links, leading to "know-how."

Professor M. Lapré, Assistant Professor of Operations Management at Owen School of Management, and L. Van Wassenhove, the Henry Ford Chaired Professor of Manufacturing at INSEAD, show that it is possible to acceler-

ate factories' learning curves through focused quality and productivity improvement efforts.

Even though a firm may be committed to increasing its
learning rates, for any number of reasons, this might be imped-
ed. For example, system details and dynamics can be highly
complex, or employees may adopt ideas and subsequent practices based on subjective interpretations and inaccurate beliefs.
The professors gathered data on nearly a decade of quality
improvement projects that took place at the factories of N.V.
Bekaert, S.A., to demonstrate how management can influence
learning efforts and, in turn, enhance learning rates.

Drawing upon Bekaert's quality improvement projects, they
analyzed each effort and scored them as either high or low in
terms of conceptual and operational learning. Then they slot-
ted each into one on the following classifications:

- *Firefighting.* Low conceptual and operational learning.
  Teams do not reflect on cause and effect relationships, obtain
  little validation, and often result in minor changes.
- *Artisan skills.* Low conceptual, high operational learning.
  The solutions work, but lack a clear reason why. High levels
  of ambiguity meant other employees did not believe the local
  skills could be applicable to other parts of the factory.
- *Nonvalidated theories.* High conceptual, low operational
  learning. The cause and effect is assessed, but there is no
  final step of verification. Quite surprisingly, these slow down
  the global rate of learning.
- *Operationally validated theories.* High conceptual and operational learning. Grounded in scientific models and statistical
  experiments, theories are developed, resulting changes are
  implemented and validated, and the solutions worked.

The only projects resulting in global learning were those
classified as operationally valid theories. Thus, the authors can
explain that although in some instances one learning type or
the other can be influential in local environments, it takes
more to accelerate the global factory learning curve.

The authors then examine the plants' special manufacturing lines created especially to accelerate the learning process (called model lines). Although the first model line consistently produced both conceptual and operational learning and yielded spectacular improvements, there were notably differing levels of effectiveness when the management tried to replicate the line in other factories. They conclude there are other factors that must be addressed, such as management buy-in and interdepartmental problem solving.

NEW ECONOMY ORGANIZATION/VIRTUAL ORGANIZATIONS/FUTURE TRENDS

The network organization.   Peter Senge, the author of *The Fifth Discipline: The Art & Practice of the Learning Organization*, is convinced that the new organization will look like a network of profit centers. He argues that more than 35 years ago, J. Forrester, an MIT professor predicted that kind of organization in a paper called "The New Corporate Design." He said there would be three forces that would bring to pass this organization as a network of profit centers: the inevitable distribution of power and authority; computer technology, which would eliminate middle management (this was written in 1965); and the understanding of complex systems, which will make it possible to develop not just one great intuitive systems thinker at the top, but many system thinkers who would manage the enterprise.

Quite independently of Forrester's ideas, R. Ackoff of Wharton came up with the same idea: break up the organization into a network of profit centers. Ackoff actually went further than Forrester in that he tried to implement it in several companies. But the network of profit centers isn't the same as decentralization. "Localization" is a better term: an organization made up of free-standing units that each have full profit-and-loss responsibilities and some synergy between them—synergy, not in the portfolio sense, but in the sense of shared learning and vision. Localization is different from a conglomerate of independent business within a holding company. Each free-standing unit has its own internal board of directors.

But these internal boards do not function as bosses. They serve as mentor and coach; they do not control the local leaders. But there are some problems around this concept of the internal boards: the skills and capabilities of the board members and the confrontation of complex issues collaboratively. The internal board structure is promising, but Senge himself thinks it takes many years to implement effectively, because there will be a never-ending battle to overcome the deeply set instinct to make the decisions for the local operating units, just as has always been done in traditional authoritarian organizations.

Disadvantages of virtuality.   The disadvantages of virtuality are:

- Operating individually while calling yourself part of a team
- Lack of or difficulty in communication
- Difficulty in managing/leading/coordinating
- Barriers against developing a sense of community or teamwork
- Weak or nonexistent culture

   How to deal with the disadvantages:

- Periodic face-to-face meetings
- Special managerial and team leadership techniques

HIGH POINTS OF "BREAKTHROUGHS IN ORGANIZATION"

- "Organization" refers to the administrative and/or functional structure of an organization. Examples of structure types: functional, process (cross- or multifunctional), matrix (functional and process combined), teams, etc.
- Successful organizations in the twenty-first century will have to become incubators of leadership and commitment. Developing that leadership and commitment demands flat

and lean structures, high levels of participation and risk-taking, and low levels of control.

- Managers need to be on guard against fads and "false trails" when establishing or changing structures.

- The key to the design and operation of organizational structures is alignment of infrastructures with strategies and resources—financial, technological, and human.

- Functional-based organizations can result in efficient operations within each function, but less than optimal results delivered to internal and external customers. There are many barriers to interfunctional communication, control, and smooth work flow across functions.

- Process-based organizations provide each process with the special functional resources necessary to accomplish the work. Barriers associated with traditional function-based organizations are eliminated, making it easier to manage processes on an ongoing basis. Reporting structures are associated with cross-functional processes, and accountability is assigned to a process owner. Process-based organizations run the risk of diminishing the skill and efficiency within functional specialties.

- Matrix-based organizations operate with a hybrid combination of the functional- and process-based structures, enjoying the advantages of each individual structure type with few of the disadvantages.

- Many managers worry about indifferent employees who show little or no commitment to the work or the company. It appears that there is a positive correlation between commitment and empowerment. The more empowered employees are, the more commitment they demonstrate.

- There is an apparently inexorable ascendance in organizations of empowered work systems such as teams and teamwork. We can expect widespread introduction of high-performance teams, self-managing teams, self-directing teams, empowered teams, natural work groups, and self-regulating teams and the like, with self-directed teams being the most common.

- Six Sigma efforts are utterly dependent on teams to provide the cross-functional synergy and perspective necessary to solve cross-functional performance problems.
- There are a number of nondelegable tasks that must be performed by management to support an empowered organization.

BREAKTHROUGHS IN CURRENT PERFORMANCE IMPROVEMENT— CLASSIC AND SIX SIGMA DMAIC

A very large teaching hospital (actually a complex of many hospitals) on the west coast of the United States found itself experiencing several quite serious problems in some of its treatment support services. The root causes of these problems were mysterious and elusive. No one seemed to really know why the problems existed or what to do about them. Many theories were offered, but no proof. The management decided that something had to be done to improve the performance of these curiously poor-performing processes. Here are the preliminary results of this decision.

- *Problem one.* "The time from when a physician orders a test for an inpatient until the time the result is reported is too long."
- *Mission one.* Reduce turnaround time for laboratory tests from the time a physician orders until the results are reported for inpatients.

- *Symptoms.* Only 50 percent of medical laboratory tests were available at the 8:30 A.M. goal.

- *Results of improvement project one.* Goal of 8:30 A.M. was met, on a sustained basis, **96 percent** of the time (up from 50 percent). This represented only the first pass; further improvements followed.

- *Problem two.* Turnaround time for testing gynecologic cytology specimens was 25 to 30 working days.

- *Mission two.* Reduce turnaround time for testing of gynecologic cytology specimens.

- *Symptoms.* Turnaround time fluctuated between 25 and 30 days.

- *Results of improvement project two.* Turnaround time decreased from 25 to 30 days to a new sustainable standard of 3 to 5 working days.

Who did what to make such dramatic—and potentially life-saving—improvements?

Simplified, this was the sequence of events:

1. Executive management took charge of the situation. It took on the nondelegable roles of a "Quality Council" whose mission was to direct and support a cross-functional organizational attack on serious quality problems, problem-by-problem, project-by-project.

   A group of prospective project team leaders and facilitators was identified and trained in quality improvement methodology—the application of the scientific method for diagnosing and curing ailing products or processes. The clinicians and medical administrators who received this training found the quality improvement approach quite congenial. The approach involves two major steps:

   a. *diagnosis* (a search for root causes,) and b. *remedy* (removing or going around the causes and putting in place new controls so the causes can't return). It closely resembles the medical approach of diagnosing and treating ailing patients.

2. Improvement project teams were formed, consisting of persons from the organizational location where the problem showed up, and persons from departments upstream from there. All teams could call on a stand-by group of trained diagnosticians. These teams received training in quality improvement methodology.

The project teams followed the steps in the quality improvement methodology to their logical and inexorable conclusion: they identified the hitherto mysterious and elusive causes of the problems, and changed the processes to make the problems go away and stay away. (See example of a quality improvement project following the description of the classic "Juran" model of the quality improvement methodology.)

BREAKTHROUGH IMPROVEMENT

Breakthrough improvement addresses the question: "How do I reduce or eliminate things that are wrong with my products, services, or processes and the associated customer dissatisfaction and high costs of poor quality (waste) that consume my bottom line?" Breakthrough improvement addresses *quality* problems—failures to meet specific needs of specific customers, internal and external. (Other types of problems are addressed by other types of breakthrough.) Quality problems almost always boil down to just a few species of things that go wrong, including:

- Excessive number of defects
- Excessive numbers of delays or excessively long time cycles
- Excessive costs of the resulting rework, scrap, late deliveries, dealing with dissatisfied customers, replacement of returned goods, loss of customers and clients, loss of goodwill, etc.

Breakthrough improvement:

- Discovers root causes of the problem

- Devises remedial changes to the "guilty" processes to remove or go around the cause(s)
- Installs new controls to prevent the return of the causes

This chapter describes the route to follow to make break-throughs in current performance.

Traversing the route requires making a number of journeys:

- *The Diagnostic Journey.* From problem to symptoms of the problem; from symptoms to theories about what may cause the symptom(s); from theories to testing of the theories; from tests to establishing root cause(s).

- *The Remedial Journey.* From root causes to remedial changes in the process to remove or go around the cause(s); from remedies to testing and proving the remedies under operating conditions; from workable remedies to dealing with resist-ance to change; from dealing with resistance to establishing new controls to hold the gains.

We will present two versions of this same model—the clas-sic "Juran" version called *quality improvement,* or *break-through,* and the more recent statistically and computer-oriented version known as *Six Sigma* or *Six Sigma DMAIC.* The model improves *both* performance and quality.

THE CLASSIC "JURAN" MODEL OF QUALITY (PERFORMANCE) IMPROVEMENT

The following is the outline of the anatomy of a quality improvement project. Because this book is written as a guide-book for managers, we will limit our detailed discussions to some of the more important activities that are carried out by management. Each of the following topics that are outlined contain a large body of technical knowledge, tools, and tech-niques. A detailed description of each outlined topic is beyond the scope of this book. Each topic is treated thoroughly elsewhere, in textbooks, workbooks, journal articles, and training materials.

Identify a project (the quality council does this)

- Nominate projects
- Evaluate projects
- Select a project
- Ask: Is it an improvement (breakthrough) project?

Establish the project (the quality council does this)

- Prepare a problem statement and a mission statement
- Select and launch a team

Diagnose the cause (the project team does this)

- Analyze symptoms
- Confirm or modify the mission
- Formulate theories
- Test theories
- Identify root cause(s)

Remedy the cause (the project team and the work group where the cause[s] originate do this, perhaps with assistance from many others who are affected by, or who contribute to, the remedy)

- Evaluate alternative remedial changes
- Design the remedy
- Design new controls on the remedy
- Design for the culture (prevent or overcome resistance to the remedial changes)
- Prove effectiveness of the remedy under operating conditions
- Implement the remedial changes

Hold the gains (the project team and the affected operating forces do this)

· Design and implement effective controls
· Foolproof the remedy, if necessary
· Audit the controls

Replicate results and nominate new projects (the quality council does this)

· Replicate the results (clone, perhaps with modifications, the remedy)
· Nominate new projects based on lessons learned from the project

QUALITY IMPROVEMENT RESPONSIBILITIES

ACTIVITIES BY MANAGEMENT	ACTIVITIES BY TEAMS
• Establish Quality Councils	• Analyze symptoms
• Select projects	• Theorize as to causes
• Write problem and mission statements	• Test theories
• Provide resources, especially time	• Establish root cause(s)
• Assign teams and projects to teams	• Stimulate remedies and controls
• Review progress	• Nominate new projects
• Provide recognition and rewards	

Here is a brief example of a straightforward, relatively simple (yet valuable) project that illustrates the classic quality improvement methodology.

THE CASE OF MYSTERIOUS DAMAGE TO LINOLEUM IN UNITS OF MANUFACTURED HOUSING

Nearly half of the residential single-family dwelling units being built in the United States are manufactured on moving lines in factories. The modular units are transported to remote locations, joined together there, and set upon prepared foundations on the home purchaser's lot. It's hard to tell the difference between an assembled manufactured house and a stick-built house once they are finished and landscaped.

A large manufacturer of modular housing units was dissatisfied with the level of very expensive rework some of its factories in various locations around the country were experiencing. Customer dissatisfaction was rising; profits were eroding. Quality Councils consisting of the general manager and all direct reports were formed at each factory. They received training in quality improvement, identified the most expensive rework, formed and trained teams in quality improvement, and set them to reducing the amount of rework. This is the story of one such improvement project.

Identify the project.   One factory's Quality Council listed and prioritized its rework problems, using the *Pareto analysis*. The Pareto distribution (arranged in descending order of cumulative percent) of their most costly rework types during the past six months looked like this:

- Replacing damaged linoleum—51 percent
- Repair of cut electrical wires in walls—15 percent
- Replacing missing fixtures at the site—14 percent
- Repairing leaks in water pipes—12 percent
- Repairing cracks in drywall—8 percent

On the basis of the Pareto analysis, the Quality Council selected public enemy number one: replacing damaged linoleum. This problem is very expensive to repair. Often walls must be removed and new linoleum laid, followed by replacing the wall.

Establish the project

- The problem statement was: "There are too many occurrences of replacing damaged linoleum."
- The mission statement was: "Reduce the number of occurrences of replacing damaged linoleum."

Note that *both* the problem and mission are described, and the variable and unit of measure in the problem statement and mission statement are identical. This is important because: a. from the problem statement, the team knows what problem it is trying to solve. The rest of the project focuses on whatever the council selects as the problem; and b. if they don't match, the team may carry out the mission and not solve the problem, if the mission statement does not match the problem statement.

The Council chartered a project team consisting of representatives of the workstations where the linoleum was installed, and where the damaged linoleum was observed. The Council appointed a worker in one of those workstations to be the project team leader. The project team leader received training not only in quality improvement, but also in leading a project team. A trained facilitator instructed and coached the team in quality improvement methodology and tools.

Diagnose the cause.   The team's first task was to analyze the symptoms. (Symptoms are outward evidence of the problem.) The primary symptom was, of course, the number of occurrences of replacing damaged linoleum. Secondary symptoms were the cost of replacement, the various types of damage, the location where the damage showed up, down time due to replacement, overtime to do replacement, and the like. The symptoms were analyzed by *defining* them, *quantifying* them, and *visualizing* them. What follows is an analysis of the primary symptom.

Various types of damage were identified and defined: gouges, scrapes, cuts, gaps, and smears. A flow diagram was constructed showing all operations in all workstations that related to linoleum or replacement of linoleum. The flow diagram also identified the workstations where damage showed up.

Several Pareto analyses were performed. The first was by type of damage. It showed:

- Gouges (dents)—45 percent
- Scrapes—30 percent
- Cuts—21 percent
- Gaps—4 percent
- Smears—2 percent

Accordingly, the team now focused, temporarily, exclusively on gouges.

A second Pareto analysis, of gouges by location in the house was performed. It showed:

- Kitchen—38 percent
- Interior hall—21 percent
- Bathroom 1—18 percent
- Bathroom 2—14 percent
- Laundry—9 percent

Now the team was focused on gouges in kitchens, to the temporary exclusion of all the other symptoms. The Pareto principle states that for any given effect (output of a process, a symptom), there are a number of contributors. These contributors make unequal contributions. A relatively few contributors make by far the greatest contribution. These are called the "vital few." Following the Pareto principle, they were going after the vital few contributors to the problem to get the greatest return for the least effort.

A third Pareto analysis, of gouges in kitchens by shift, showed no difference in occurrences between shifts, indicating that "shift" is not a contributor to gouges in kitchens.

The team now theorized about what could cause gouges in kitchens. They generated a long list of theories. The most compelling ones were:

- Dropping heavy, sharp objects (tools)
- Dragging objects across the floor
- Grit on boots
- Poor or no training of employees
- No protection of new linoleum

In the manufactured housing industry, it is known that the first three theories, if they in fact occur, do cause gouges in

linoleum. Those theories did not need testing. What about
"lack of protection?" The only way to test that theory is first to
correlate gouges in kitchens with the presence or absence of
protection. The team arranged that all reports of linoleum
damage or replacement would include an indication that protection was "present."

When this was done, it was discovered that virtually all
cases of gouge damage to linoleum occurred when floor protection was missing. Furthermore, the team discovered that
there was no formal control plan for protection to be installed.
Consequently, no quality inspections or quality assurance
checks revealed that no controls were being exercised and that
none even existed! Floor protection was a haphazard phenomenon at best.

Remedy the cause.   Workers, purchasing personnel, and
engineers went to work to select and procure material that was
strong and economical to lay on freshly installed linoleum. All
agreed that the operator would be responsible for laying it
immediately after each job, and that supervisors would check
to see that it happened. Incidents of gouge damage—and other
types of damage—to linoleum went down dramatically. (It
seemed several damage types had common causes, one of
which was no protection.) For a few weeks, damage to linoleum
almost entirely disappeared. Celebrations were held. The Plant
Manager began to look forward to granting bigger bonuses—
and getting one himself!

At the weekly meeting of the factory management team a
few weeks later, the Quality Manager reported the mysterious
reappearance of gouge damage. This news was greeted with
incredulity and disappointment. "We thought we had gouge
damage licked!" And indeed they had, except for a couple of
"small details."

Hold the gains.   When the team investigated, it discovered
that 1) no formal control plan for proving protection had been
devised and published; and 2) there had been turnover of workers in the various workstations who had not been trained in the
procedure; 3) the new workers had not been trained because

there was no published plan, and what is more, there also was no formal training program (with controls to assure that training actually happened). Consequently, no training either could or did take place. It became apparent that the "factory" operated more like a construction site under a roof, with standards upheld by the skill and pride of craftsmen. A factory, by contrast, is characterized by more formal procedures and controls. All this was a valuable lesson learned for all concerned, and led to a number of additional new improvement and planning projects, new attitudes toward the work, and a maturing of the plant as it evolved from construction site to factory. Controls and training were formalized.

THE SIX SIGMA MODEL OF QUALITY (PERFORMANCE) IMPROVEMENT

Define (the champions and executive council do this)

- Identify potential projects
- Evaluate projects
- Select projects
- Prepare a problem and a mission statement, and a team charter
- Select and launch teams

Measure (the project team does this)

- Measure baseline performance
- Map and measure the process creating the problem
- Plan for data collection to:
  a. Measure key product characteristics (outputs, Ys) and process parameters (inputs, Xs)
  b. Measure key customer requirements (CTQs—"critical to quality")
  c. Measure potential failure modes
  d. Measure the capability of the measurement system
  e. Measure the short-term capability of the process

Analyze (the project team does this)

· Analyze response variables (outputs, Ys)

· Analyze input variables (Xs)

· Analyze relationships between specific Ys and Xs, especially cause–effect relationships

· Confirm determinants of process performance (vital few Xs)

Improve (the project team, often with help of others, does this)

· Plan designed experiments

· Conduct screening experiments to identify the critical, vital few process determinants (Xs)

· Conduct designed experiments to establish a mathematical model of process performance

· Optimize process performance

· Evaluate possible improvements

· Design and implement the improvements

Control (the project team and the operating forces do this)

· Design controls and document improved process

· Validate the measurement system to be used in controls

· Establish process capability of improved process

· Implement new process and monitor it

Many quality or performance problems can be solved using either method. The primary enhancement to the quality improvement process found in the Six Sigma process is the use of laptops and statistical software packages. If your problems are elusive, or if you must attain quality levels measured in parts per million, and approaching perfection, Six Sigma will place your ailing process under a microscope of unprecedented precision and clarity, and make it possible to understand and control the relationships between input variables and desired output variables.

You have a choice of what "weapon system" to bring to bear on your problems: a "conventional" weapon system (quality improvement) or a "nuclear" system (Six Sigma.) The conventional system is perfectly effective on many problems and much cheaper than the more elaborate and demanding nuclear system. The return on investment is considerable from both approaches, but especially so from Six Sigma if your customers are demanding maximum quality levels.

The following pages contain a basic overview of the Six Sigma improvement process, the so-called "DMAIC" process. DMAIC stands for: define-measure-analyze-improve-control.

There are two variations of this model: one that addresses repetitive *production* processes and one that addresses repetitive *transactional* processes. It is beyond the scope of this managerial-oriented book to describe the distinctions in detail. Suffice it to say that if you are confronted with problems in both areas, you should know that there is a preferred approach for the type of problem your organization wishes to solve.

DMAIC PRODUCTION AND TRANSACTIONAL PROCESSES

We need to clearly establish the difference between production (a.k.a., manufacturing) and transactional processes.

All processes are transformations that result in the change of state of one or more things that can be physical objects or services.

Production processes directly transform raw materials or semi-finished goods into a final physical product (a.k.a., good). Examples of production processes include:

- Chemical processes (distilling, reducing, etc.)
- Mixing (with or without change of phase)
- Forming
- Changing properties (heat treatment, melting, casting, molding, surface treatment, etc.)

- Finishing
- Assembly
- Testing
- Packing

The output of production processes is a transformed physical product; these processes are deterministic, workflow-oriented, highly procedural, and therefore highly repeatable. Because of this, production processes are well suited for representation by the traditional, workflow-based input-process-output (IPO), or supplier-input-process-output-customer (SIPOC) models.

The definition:

> A production process is a series of work activities performed by people and other resource-consuming assets in order to transform given input(s) into output(s).

Transactional processes (sometimes also called *people or paper processes*) directly transform one state or condition of one or more things (objects, abstractions such as information, data, symbolic representations, etc.) into another. One execution of a transactional process results in a transaction, the outcome of which, in turn, may be a change of state in a number of things (physical objects such as inventories, data and information, people, etc.). Examples of transactional processes include:

- Value-added service processes related to production (transporting, installing, storing, repairing, maintaining, etc.)
- Support or back-office processes in manufacturing and service organizations (selling, purchasing, subcontracting, warehousing, billing, human resources, etc.)
- Value-added processes in service industries (banking, insurance, transportation, health care, hospitality, education, etc.)
- Value-added processes in the public sector (including the military) and the not-for-profit sector (legislative and administrative processes, planning, command and control, fund raising, etc.)

The output of transactional processes is a change of state or condition, defined by the transaction. These processes are information (communication) driven in that successive executions of a transactional process depend on the informational inputs (requests, offers, etc.) received at the outset of each execution. Accordingly, successive executions may be different with different results. Therefore, these processes are not always repeatable, but are self-regulating and highly adaptable.

The definition:

A transactional process is a logical set of customer–supplier tasks that drive work activities performed by people.

Transactional process characteristics that differentiate them from production processes include:

- Scarcity of measurement data; available measurements are primarily discrete (attribute)
- Measurement system is partially or entirely I/T-defined (e.g., reporting)
- The definition of quality includes information quality
- Dominant variables: people and information
- High-cost labor
- Disproportionately large financial leverage

SIX SIGMA DMAIC

DMAIC: DEFINE

The steps you will find in the Define phase in the following pages are virtually the same as those used in the classic quality improvement approach. The rest of the phases look somewhat different, even though both quality improvement and DMAIC follow the same problem-solving steps.

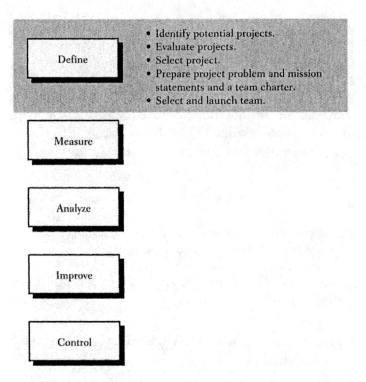

OVERVIEW OF THE DEFINE PHASE

In the Define phase, potential Six Sigma projects are identified. Nominations can come from various sources, including customers, reports, and employees. To avoid suboptimization, management has to evaluate and select the projects. While evaluation criteria for project selection are many, the primary basis should be the costs of poorly performing processes (COP^3) at the company or division level.

The project problem and mission statements, as well as a team charter, are then prepared and later confirmed by management. Management selects the most appropriate team of personnel for the project and assigns the necessary priority. Project progress is monitored to ensure success.

Define: Deliverables

· List of potential projects
· COP^3, ROI, and contribution to strategic business objective(s) for each potential project

- Selected project(s)
- Project problem statement and mission statement for each project
- Formal project team(s) headed by black belt

Questions to Be Answered

- What customer-related issues confront us
- What mysterious, costly quality problems do we have that should be solved?
- What are the likely benefits to be reaped by solving each of these problems?
- Which of our list of problems deserves to be tackled first, second, etc.?
- What formal problem statement and mission statement should we assign to each project team?
- Who should be the project team members and leader (black belt) for each project?

1. Search various sources of information (customers, reviews/audits, business plans, employees) to discover which defects occur most frequently.
2. Analyze costs of poor quality to discover which defects are most expensive and wasteful.
3. Investigate prime sources of information about the organization, including:
   - *Customers.* Both customer complaints and customer opinion
   - *Reviews/audits.* Data on the costs of poor performance, score cards, internal audits, and quality system audits
   - *Business plans.* Long-term strategic goals, business plans, or other business objectives
   - *Other improvement projects.* Existing projects may be separated into smaller, more manageable components or, while solving one problem, new problems can be uncovered

- *Employees*. The organization's managers and other employees are often the first to recognize opportunities for improvement
4. List potential projects from each source. At this point, quality problems should be listed as potential projects. This is not the time to judge the projects. The objective is to obtain as comprehensive a list as possible.

COP³, ROI, and contributions to strategic business objectives

1. Using performance and COP³ data, perform multiple Pareto analyses to discuss which potential project is most costly.
2. Using Pareto and COP³ analysis, selection matrices, and referring to the strategic objectives in the business plan, identify the projects that should make the greatest impact on the business.
3. After projects are nominated, each needs to be evaluated objectively in terms of potential impact on:
   - Retaining customers
   - Attracting new customers
   - Reducing the costs of poorly performing processes
   - Enhancing employee satisfaction
4. To evaluate possible projects, we require data on:
   - The sources of complaints and dissatisfaction most likely to drive away existing or new customers
   - The competition's level of performance compared with ours
   - The most costly deficiencies
   - The deficiencies in internal processes that have the most adverse effect on employees
5. Specific, objective data are needed on at least some of the above four topics.
6. Using data is essential for two reasons:

- Data tell us which problems are the most important
- Only with data can we know whether the project has brought about any improvement

Example. One paper mill set out to identify the most significant breakthrough improvement projects by examining the costs of poorly performing processes quality associated with its production process. The following diagram (Figure 8.1, called a Pareto diagram) summarizes its results.

Each bar in the diagram represents the costs associated with one category of poor quality. The line drawn above the bars documents the cumulative percent of the total costs for all the categories to the left of the point on the line. (For example, the first two categories—"broke" and "claims"—together account for approximately 75 percent of the total costs of poor quality.)

Of the seven types of unnecessary quality costs that this paper mill experienced, by far the highest is so defective that it

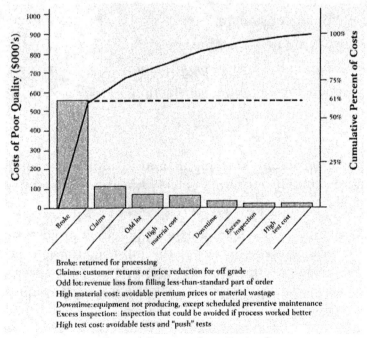

Broke: returned for processing
Claims: customer returns or price reduction for off grade
Odd lot: revenue loss from filling less-than-standard part of order
High material cost: avoidable premium prices or material wastage
Downtime: equipment not producing, except scheduled preventive maintenance
Excess inspection: inspection that could be avoided if process worked better
High test cost: avoidable tests and "push" tests

FIGURE 8.1   Paper mill costs of poor quality—Pareto diagram.

was returned for reprocessing. It was treated with chemicals, added to the other raw materials, and run all the way through the papermaking machinery again. This one cost accounted for 61 percent of the total. To reduce costs of poor quality, a project was clearly needed to address the problem of "broke."

7. Collect specific, objective data on each problem identified as a potential project. The data should answer these questions:
   - What complaints and dissatisfaction are most likely to drive away existing or new customers?
   - What level of quality does the competition deliver and how does it compare with ours?
   - What are our most costly deficiencies?
   - Which deficiencies in our internal processes have the most adverse effect on employees?

8. Use the data to determine each potential project's probable impact on:
   - Retaining customers
   - Attracting new customers
   - Reducing the cost of poor quality
   - Enhancing employee satisfaction
   - Achieving business objectives

Assign a problem statement and a mission statement. Using evaluation criteria, select specific project(s)—each assigned a specific problem and mission.

Select project.   Reviewing data on potential projects against specific criteria helps to select the most appropriate project. The following criteria Table 8.1 lists and explains seven criteria for selecting a project and includes appropriate questions for each one.

TABLE 8.1   Criteria for Selecting a Project

CRITERIA	QUESTIONS
Chronic	Is the problem chronic? The project should correct a continuing problem, not a recent specific episode.
Significance	How significant do you expect the results to be? When a project is completed, significant favorable results should be evident internally, and by the customer. The results should be worth the effort.
Size/scope	Is the project a manageable size? Most quality improvement projects should take less than a year to complete. Many, in fact, can be completed in less than six months. If it appears a project will take a long time to complete, it is usually possible to divide it into smaller projects likely to yield results more quickly.
Measure of potential impact	What is the project's potential impact? Impact must be measured. Typical measures include the project's potential to: • Retain customers and attract new ones. • Reduce the costs of poor quality. • Provide return on investment. • Enhance customer satisfaction. • Enhance employee satisfaction.
Urgency	How urgent is the project to the organization? A project may be urgent if it addresses quality problems in core services, problems that make the organization highly vulnerable to the competition, or issues that are crucial to key customers. Problems in these areas are usually critical and should be corrected promptly.
Risk	What are the risks? If there are known or suspected risks, a project is likely to take a long time to complete or have an uncertain outcome. This does not mean the project should be avoided. It simply means the expected payoffs should be high. Projects are likely to be risky if there is reason to believe that they might involve new or unproven technology or affect departments that plan or have recently undergone major organizational changes.
Potential resistance to change	What kinds of resistance might the project create? Any quality improvement project causes change, and change frequently causes some resistance. The source of resistance may be a difficult manager whose input is important, or an entrenched organizational culture, tradition, or policy. When the choice is among several projects of equal duration, impact, significance, size, urgency, and risk, it is usually best to select the project likely to meet the least resistance.

In organizations relatively new to breakthrough improvement, there are two additional criteria to consider when selecting a project. These are shown in Table 8.2.

TABLE 8.2   Two Additional Criteria

CRITERIA	QUESTIONS
The project should be a sure winner.	Is the project a sure winner? The first projects for an organization new to quality management provide opportunities to learn and adapt the quality improvement process. For this reason, there should be no obstacles in the way of successful outcomes. First projects should still address chronic, significant problems, but not necessarily the most significant ones. It is especially important to keep early projects bite-sized and to avoid substantial resistance to change. Potential impact and urgency are of secondary importance.
The problem must be measurable.	Is the problem measurable? All quality improvement projects require measurable problems, but sometimes organizations do not yet have solid data with which to evaluate the potential impact of first projects. Nevertheless, no project should be undertaken if the problem cannot be measured. If no data exist, the project team will need to develop that data during its early work.

IMPROVEMENT: HOW TO EAT AN ELEPHANT

Occasionally, breakthrough improvement projects appear to be "elephant-sized;" that is, they cover so broad an area of activity that they must be subdivided into multiple "bite-sized" projects. When this occurs, one project team can be assigned to "cut up the elephant." Other teams are then assigned to tackle the resulting bite-sized projects. Unless such subdivision takes place, the original team may disband in frustration or take years trying to carry out the broad project.

Evaluate a project

1. Evaluate each project to validate that it is:
   - Chronic
   - Significant, with financial impact

- Manageable in size
- Likely to be a winner (if the organization is new to quality improvement)
- Able to be measured. All successful improvement projects are measurable problems

2. Rate the relative strength of each potential project with respect to:
   - Measuring potential impact to retain customers and attract new ones
   - Reducing the costs of poorly performing processes
   - Providing return on investment
   - Enhancing customer satisfaction
   - Enhancing employee satisfaction
   - Urgency
   - Risks
   - Potential resistance to change

3. Select best project(s).

4. Ask: Is it a candidate for breakthrough improvement?

   The breakthrough improvement process (DMAIC) is not a suitable approach to every organizational problem.

   A breakthrough improvement project is a structured approach to identifying and removing the root causes of performance problems in an existing process. To be certain that breakthrough improvement methods are an appropriate approach for your proposed project, ask the following questions. If the answer to each question is yes, you probably have a breakthrough improvement (DMAIC) project.

5. Are we trying to reach a new level of performance for an existing good or service?

6. Have we measured (or could we measure) specific deficiencies or opportunities for performance improvements?

7. Are we trying to find and eliminate the root cause of a problem?

If you answer yes to the following question, however, you probably do *not* have an improvement project.

8. Do we want to develop a brand-new product or process?

A "yes" answer suggests the need for a Design for Six Sigma (DFSS) project, not a Six Sigma breakthrough improvement (DMAIC) project.

PREPARE A PROBLEM STATEMENT AND A MISSION STATEMENT

Preparing the problem and mission statements are the first order of business for a company when establishing a project. These statements are the written instructions to the team selected to tackle a Six Sigma breakthrough improvement project.

1. Write a (usually one-sentence) description of the problem to be solved by each team—stated in terms of what is wrong.

2. Write a (usually one-sentence) description of each team's mission (what the team is to accomplish with respect to the problem).

Note: Most quality or performance problems distill down to one or more of three variables: costs, cycle times, frequency of defects. The variables and units of measure described by the problem statement should be the very same variables and units of measure that are described in the mission statement.

The problem and mission statements should make it possible for the team to understand what problem they are trying to solve and what they are empowered to do about it.

CRITERIA FOR DESCRIBING THE PROBLEM

An effective problem statement is:

• *Specific.* It explains exactly what is wrong and distinguishes the deficiency from similar problems.

- *Observable.* It describes visible evidence of the problem.

- *Measurable.* It indicates the scope of the problem in quantifiable terms by answering, "How much?" "How many?" or "How often?" Measurement is important for two reasons. First, it helps to determine whether the problem is large enough to justify attention. Second, if the project goes forward, it provides criteria for evaluating the remedy. If no measurements exist, they should be developed by the breakthrough improvement team before the team begins to diagnose the root cause.

- *Manageable.* A manageable problem is one that can probably be solved in 6 to 12 months. If a problem is too large, it should be broken down into several smaller, more manageable projects.

Example.   Our company's procedure for shipping replacement parts takes 10 days longer on average than our major competitors take.
    The statement is:

- *Specific.* It names a particular process and what the problem is.
- *Observable.* Evidence of the problem can be obtained from internal reports and customer feedback.
- *Manageable.* The problem is limited to one type of shipping procedure.

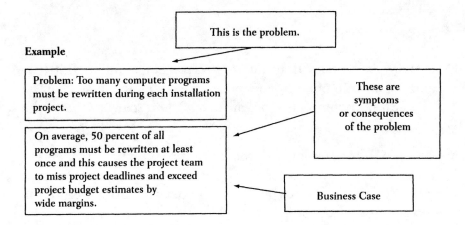

CRITERIA FOR DESCRIBING THE MISSION

An effective mission statement indicates the objective of the project, that is, what the project team is to do about the problem.

Example. Problem: Our company's procedure for shipping replacement parts takes 10 days longer, on average, than our major competitors take.

Mission: Reduce the shipping time for replacement parts to less than or equal to that of our major competitors.

> **Note:** The mission statement should contain the same variable and units of measure as does the problem statement. If the variable and units of measure are not the same, then the mission (or objective) doesn't match the problem; and even if carried out, it may not solve the problem. In the preceding example, the problem is expressed in terms of length of shipping time (number of days). So, the objective is also expressed as "length of shipping time" (number of days).

PITFALLS TO AVOID

A mission statement should not:

- Imply a cause
- Suggest a remedy
- Assign blame

Implying a cause. It is up to the project team to find the cause of a problem. Preconceived ideas about cause may be inaccurate, incomplete, or mistaken, and can mislead the team.

Example. Our company's procedure for shipping replacement parts takes 10 days longer, on average, than our major competitors take. Eliminate the delays in documentation before the replacement part is shipped.

After collecting and analyzing data, it might turn out that warehousing procedures were the chief cause of the problem.

Suggesting a remedy. Without knowing the cause, it is not possible to find an effective improvement. Attempts to solve a problem without knowing the cause are doomed to failure.

Example. Our company's procedure for shipping replacement parts takes 10 days longer, on average, than our major competitors. Install a computerized shipping log to speed up the process.

The mission implies that the procedure would be improved with a computerized shipping log no matter what the actual cause of the delay.

- Team charters (optional)—Beside a problem statement and a mission statement, many teams also receive a charter from management detailing specific team responsibilities and giving the team the authority to carry out its mission. Typical provisions of a team charter are described below.

- The team is expected to apply the steps of the Six Sigma breakthrough improvement process throughout the project. These are often listed explicitly.

- The team is authorized to collect relevant data, discuss the problem with those involved with it, and develop an improvement.

- Team members are asked to spend a specific amount of time on the project each week—including both time in team meetings and time between meetings for preparation, data collection, etc. (typically a total of four hours, but often more).

- The team has access to the resources it needs to carry out the mission.

- Arrangements are spelled out for obtaining additional resources and policy support if the project proves to be beyond the resources already at the team's disposal.

Form a Team

1. Assign a cross-functional team to each project including people:

- From where in the organization the problem shows up (where the pain is observed or felt)
- From where in the organization the sources or courses of the problem likely will be found
- Who have high levels of diagnostic skills
- Who may be helpful in implementing the remedy

2. Evaluate each team member and the team as a whole. Each team member should have:

- Direct, detailed, personal knowledge of some part of the problem
- Time for team meetings and assignments

3. As a group, the team should be able to:

- Describe major elements of the process
- Explain how the parts of the process relate to one another
- Work with their departments to implement the remedy

Before continuing the overview of the MAIC phases, we should point out a basic principle of Six Sigma: The output (Y) of a process is dependent on the inputs (Xs) to the process. What comes out of a process is a result of what goes in. We can say that an output (Y) is a function (f) of inputs (Xs). Stated mathematically, this reads $Y=f(X)$, or, more precisely, $Y=f(X1+X2+...Xn)$. The job of a Six Sigma improvement team is to discover the Xs (the inputs, the causes) of a serious performance problem (a "bad" Y), remove or change the Xs, and put into place new controls so the original Xs and the Y can't return.

DMAIC: MEASURE

OVERVIEW

The project team begins process characterization by measuring baseline performance (and problems) and documenting the process, as follows:

1. Map the process

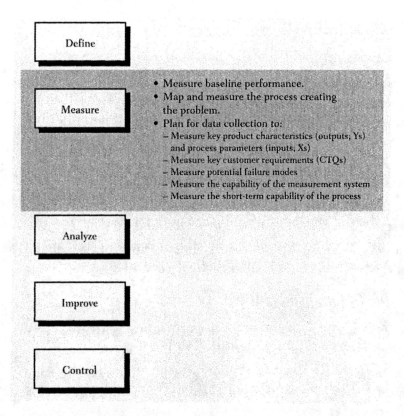

2. Identify key customer requirements (CTQs)
3. Determine key product characteristics and process parameters
4. Identify and document potential failure modes, effects, and criticality

The purpose is to identify and document the process parameters (or input variables, Xs) that affect process performance and product characteristics (or output variables, Ys) of critical interest to the customer. Process documents are updated as the project progresses. The team then plans for data collection for the rest of the measure phase and the next (analyze) phase. Then, the measurement system is validated, and process capability is measured.

Measure: Deliverables

- Baseline performance metrics describing outputs (Ys)
- Process flow diagram; key process input variables (KPIVs); key process output variables (KPOVs); cause–effect diagram; function deployment matrix (FDM)—optional; potential failure mode and effect analysis (FMEA) (to get clues to possible causes [Xs] of the defective outputs [Ys])
- Data collection plan, including sampling plan
- Minitab Gage R&R Six Pack or Attribute GR&R Effectiveness— to describe the capability of the measurement system(s)
- Capability Six Pack—to describe short-term process capability and determine extent of statistical control

Questions to be answered

- How well is the current process performing with respect to the specific Ys (outputs) identified to Pareto analyses?
- What data do we need to create in order to assess the capability of (a) the measurement system(s) and (b) the production process(es)?
- What is the capability of the measurement system(s)?
- Is the process in statistical control?
- What is the capability of the process(es)?

Determine baseline performance. Measure the actual performance (outputs; Ys) such as costs of poor quality, number of defects, cycle times of the process(es), which creates the problem to discover—by Pareto analysis—which vital outputs (Ys) make the greatest contribution to the problem.

Map the process. Focusing on the vital one (or few) outputs (Ys) identified by the Pareto analysis, graphically depict the process that creates it (them) by mapping the process with a flow diagram in order to understand the process anatomy.

Identify key input and output variables. To understand the process in more detail, analyze the flow diagram to identify key process input variables (KPIVs; Xs) and key process output variables (KPOVs; Ys) associated with each process step, and indicate which of these process steps add value and which don't.

Focus on most significant variables. In order to narrow further the focus of the project team, and to prioritize which specific KPIVs (Xs) and KPOVs (Ys) the team will examine, create a functional deployment matrix (FDM)—utilizing the list of KPIVs and KPOVs generated in the process analysis.

The FDM will identify the KPIVs and KPOVs that have the greatest impact on key customer requirements. It will also translate customer requirements into product design specifications (desired Ys) and, in turn, translate design specifications into appropriate part, process, and production requirements (desired Xs, CTQs).

Measure potential failure modes. Referring to the analyzed process flow diagram (PFD) and the FDM, for each process step, perform a failure mode and effect analysis (FMEA) by listing potential process defects (Ys) that could occur, their effects (Ys; KPOVs) and their potential causes (KPIVs, Xs). (An additional source of ideas of possible KPIVs is the cause–effect diagram, which displays brainstormed possible causes for a given effect.) In addition, rate the severity of each effect, the likelihood of its occurrence, and the likelihood of its being detected should it occur. Upon completing the analysis, you will be able to identify those potential process failures that have the most risk associated with them. These results are used to further focus the project on those variables most in need of improvement.

Plan data collection for short-term capability study

1. In preparation for determining the capability of the measurement system (which measures the KPIVs and KPOVs upon which the project team has focused) and the short-term capability of the process, create a sampling and data collection plan.

2. In preparation for determining the short-term capability of the process, determine the capability of the measurement system (which will be used to measure KPIVs and KPOVs) to provide consistently accurate and precise data upon which the project team can depend to "tell the truth" about the process.

3. If the measurement system is found to be not capable, take corrective action to make it capable.

4. If the measurement system is found to be capable, proceed with the next step—determining if the process is in statistical control with respect to given variables (Ys).

Measure the short-term capability of the process

1. In preparation for measuring short-term capability of the process to meet given specifications (Ys), ascertain whether the process is in statistical control with respect to the given output (Y) of interest.

2. If the process is not in statistical control, that is, if control charts detect special causes of variation in the process, take action to remove the special causes of variation before proceeding with the process mission.

3. If the process is in statistical control, that is, the control charts do not detect special causes of variation in the process, perform a short-term capability study to provide baseline data of the ability of the process to consistently produce a given output (Y).

Confirm or modify the mission

1. Evaluate the project's problem statement and mission statement.

   • Does the problem statement and the mission statement meet all the criteria of an effective problem statement and mission statement, and have clearly defined boundaries?

   • Are the same variables and units of measure found in the problem statement also found in the mission statement?

- Can the project be handled by a single team?
- Does it avoid unnecessary constraints, but still specify clearly any necessary global constraints such as organizational strategy?
- Are there any points that need clarification?
- Are the team members representative of departments, divisions, or work units affected by the project?

2. Verify that the problem exists. If the problem has not been measured, the team must do so at this point.

3. Validate project goal(s). Verify that the basis for the project goal(s) is (are) from one or more of the following:
   - Technology
   - Market
   - Benchmarking
   - History

4. Modify the problem statement and mission statement if either do not meet the criteria as described in tasks 1 and 2.

5. Obtain confirmation from the leadership team, champion, black belt, or quality council on any necessary changes to the project mission or to team membership.

6. Create a glossary for your project that will serve as a "dictionary" for all important terms relating to your project. Select a team member to act as glossary chief with the responsibility of maintaining the project glossary.

DMAIC: ANALYZE

OVERVIEW OF THE ANALYZE PHASE

In the Analyze phase, the project team analyzes past and current performance data. Key information questions formulated in the previous phase are answered through this analysis. Hypotheses on possible cause–effect relationships are developed and tested. Appropriate statistical tools and techniques are used: histograms, box plots, multi-vari analysis, correlation and regres-

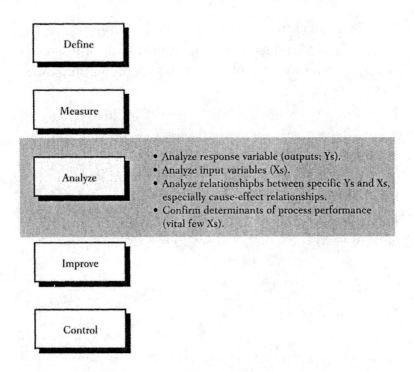

Define

Measure

Analyze
- Analyze response variable (outputs; Ys).
- Analyze input variables (Xs).
- Analyze relationshipbs between specific Ys and Xs, especially cause-effect relationships.
- Confirm determinants of process performance (vital few Xs).

Improve

Control

sion, hypothesis testing, contingency tables, and analysis of variance (ANOVA). In this way, the team confirms the determinants of process performance (i.e., the key or "vital few" inputs that affect response variable[s] of interest are identified).

It is possible that the team may not have to carry out designed experiments (DOE) in the next (Improve) phase if the exact cause-effect relationships can be established by analyzing past and current performance data.

Analyze: Deliverables

- Histograms, box plots, multi-vari study results, correlation and regression analyses—to analyze response variables (Ys)
- Results of hypothesis testing—to analyze input variables (Xs)
- List of vital few process inputs (Xs)

Questions to Be Answered

- What patterns, if any, are demonstrated by current process outputs (Ys) of interest to the project team?

- What process inputs (Xs) seem to determine each of the outputs (Ys)?
- What are the vital few Xs on which the project team should focus?

Analyze response variables (outputs, Ys) and input variables (Xs)

1. Perform multi-vari analysis to:
   - Visually narrow the list of important categorically discrete input variables (Xs)
   - Learn the effects of categorically discrete inputs (Xs) on variable outputs (Ys) and display the effects graphically

2. Perform correlation and regression to:
   - Narrow the list of important continuous input variables (Xs) specifically to learn the "strength of association" between a specific variable input (Xs) and a specific variable output (Ys)

3. Calculate confidence intervals to:
   - Learn the range of values that, with a given probability, include the true value of our estimated population's parameter, which has been calculated from a sample (e.g., the population's center and/or spread)
   - Analyze relationships between specific Ys and Xs, especially cause–effect relationships
   - Confirm vital few determinants (Xs) of process performance (Ys)

4. Perform hypothesis testing using variables data to:
   - Answer the question, "Is our population actual standard deviation the same as or different from its target standard deviation?" Perform χ^2 tests.
   - Answer the question, "Is our population actual mean the same as or different from its target mean?" If sample size is more than 30, perform 1-sample Z-tests. If sample size is less than 30, perform 1-sample t-tests.

- Answer the question, "Is our population mean the same or different after a given treatment as it was before the treatment? Perform paired t-tests.
- Answer the question, "Are several (>2) means the same or different?" Perform analysis of variance.

5. Perform hypothesis testing using attribute data to:
   - Answer the question, "Is the proportion of some factor in our sample the same or different from the target proportion?" Perform minitab test and calculation of confidence interval for one proportion.
   - Answer the question, "Is proportion$_1$ the same or different from proportion$_2$?" Perform minitab test and calculation of confidence interval for two proportions.
   - Answer the question, "Is a given output (Y) independent of, or dependent on, a particular input (X)?" (This involves testing the theory that a given X is an important casual factor that should be included in our list of vital few Xs.)

DMAIC: IMPROVE

OVERVIEW OF THE IMPROVE PHASE

In the Improve phase, the project team seeks to determine the cause–effect relationship (mathematical relationship between input variables and the response variable of interest) so that process performance can be predicted, improved, and optimized. The team plans the designed experiments (DOE). Screening experiments (fractional factorial designs) are used to identify the critical or "vital few" causes or determinants. A mathematical model of process performance is then established using 2^k factorial experiments. If necessary, full factorial experiments are carried out. The operational range of input or process parameter settings are hence determined. The team can further fine-tune or optimize process performance by using such techniques as response surface methods (RSM) and evolutionary operation (EVOP).

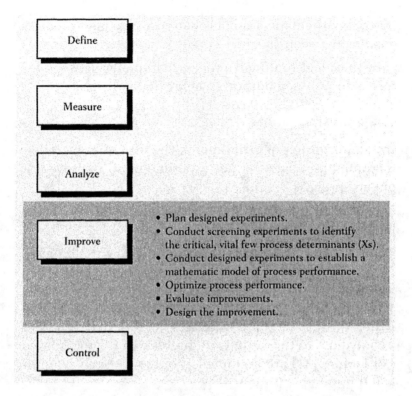

Improve: Deliverables

- Plan for designed experiments
- Reduced list of vital few inputs (Xs)
- Mathematical prediction model(s)
- Established process parameter settings
- Designed improvements

Questions to be answered

- What specific experiments should be conducted to arrive ultimately at the discovery of what the optional process parameter settings should be?
- What are the vital few inputs (Xs, narrowed down still further by experimentation) that have the greatest impact on the outputs (Ys) of interest?

- What is the mathematical model that describes and predicts relationships between specific Xs and Ys?
- What are the ideal (optimal) process parameter settings for the process to produce output(s) at Six Sigma levels?

Plan designed experiments

1. Learn about design of experiments (DOEs) in preparation for planning and carrying out experiments to improve the "problem" process.

2. In preparation for designing factorial experiments, learn about randomized block design.

3. Design, in detail, the experiments required by the project.

Conduct fractional factorial screening experiments. Perform fractional factorial screening experiments to reduce even further the list of input variables to the vital few that strongly contribute to the outputs of interest. (A relatively large number of factors [Xs] are examined at only two levels in a relatively small number of runs.)

Conduct further experiments, if necessary

1. Perform 2k factorial experiments. Multiple factors (Xs, identified by screening experiments) are examined at only two levels to obtain information economically with relatively few experimental runs. Precise mathematical relationships between Xs and Y are discovered by constructing equations that predict the effect on output Y of a given causal factor X. In addition, not only are the critical factors (X) identified, but also the level at which each factor performs the best.

2. If necessary, perform full factorial experiments. More information than is provided by 2^k factorial experiments may be required. A full factorial experiment produces the same type of information as a 2^k factorial does, but does so by examining multiple factors (Xs) at multiple levels.

Establish mathematical models of process performance

1. Using results of experiments, establish optimal settings for process parameters (Xs) to achieve desired Ys.

2. If necessary, and in addition, utilize response surface methodology (RSM) and/or evolutionary operations (EVOPs) techniques to further assist in determining optimal process parameters.

3. Using results of experiments, establish optimal settings for process parameters (Xs) to achieve desired (Ys).

Evaluate and choose optimal improvements

1. Identify a broad range of possible improvements.

2. Agree on criteria against which to evaluate the improvements and on the relative weight each criterion will have. The following criteria are commonly used.
   - Total cost
   - Impact on the problem
   - Benefit–cost relationship
   - Cultural impact or resistance to change
   - Implementation time
   - Risk
   - Health, safety, and the environment

3. Evaluate the improvements using agreed upon criteria.

4. Agree on the most suitable improvements.

Design the improvements

1. Evaluate the improvements against the project mission. Verify that it will meet project goals.

2. Identify the following customers of the improvements:
   - Those who will create part of the improvements
   - Those who will operate the revised process
   - Those served by the improvements

3. Determine customer needs with respect to the improvements.

4. Determine the following required resources:
   - People
   - Money
   - Time
   - Materials

5. Specify the procedures and other changes required.

6. Assess human resource requirements, especially training.

7. Verify that the design of the improvement meets customer needs.

DMAIC: CONTROL

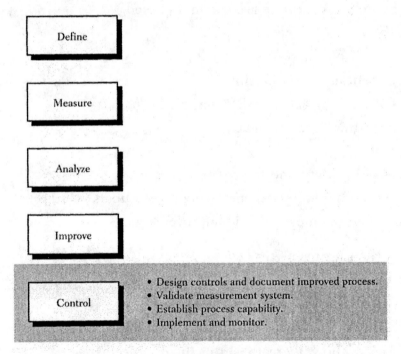

OVERVIEW OF THE CONTROL PHASE

The project team designs and documents the necessary controls to ensure that gains from the Six Sigma improvement effort can be held once the changes are implemented. Sound quality principles and techniques are used, including the concepts of self-control and dominance, the feedback loop, mistake-proofing, and statistical process control. Process documentations are updated (e.g., the failure mode and effects analysis), and process control plans are developed. Standard operating procedures (SOP) and work instructions are revised accordingly. The measurement system is validated and the improved process capability is established. Implementation is monitored, and process performance is audited over a period of time to ensure that the gains are held. The project team reports mission accomplished to management, and upon approval, turns the process totally to the operating forces and disbands.

Control: Deliverables

- Updated FMEA, process control plans, and standard operating procedures
- Validated capable measurement system(s)
- Production process in statistical control and capable to get as close to Six Sigma levels as is optimally achievable
- Updated project documentation, final project reports, and periodic audits to monitor success and hold the gains

Questions to be answered

- What should be the plan to assure the process remains in statistical control and produces defects only at or near Six Sigma levels?
- Is our measurement system capable of providing accurate and precise data with which to manage the process?
- Is our new process capable of at or near Six Sigma levels?
- How do we assure that all people who have a role in the process are in a state of self-control (have all the means to be successful on the job)?
- What standard procedures should be in place, and followed, to hold the gains?

Design controls and document improved process

1. Update FMEA, to assure no necessary controls have been overlooked.

2. Mistake-proof the improvement(s), if possible.

   [a] Identify the kind(s) of tactic(s) that can be incorporated into the improvements to make it mistake-proof. Some options include:

   - Designing systems to reduce the likelihood of error

   - Using technology rather than human sensing

   - Using active rather than passive checking

   - Keeping feedback loops as short as possible

   [b] Design and incorporate the specific steps to mistake-proof as part of the improvements.

3. Design process quality controls to assure that your improved levels of inputs (Xs) and outputs (Ys) are achieved continuously; and place all persons who will have roles in your improved process into a state of self-control, to assure that these persons have all the means necessary to be continuously successful.

   [a] Provide the means to measure the results of the new process:

   Control subjects
   - Output measures (Ys)

   - Input measures (Xs)

   [b] Establish the control standard for each control subject.

   - Base each control standard on the actual performance of the new process

   [c] Determine how actual performance will be compared to the standard.

   [d] Design actions to regulate performance if it does not meet the standard. Use a control spreadsheet to develop an action plan for each control subject.

   [e] Establish self-control for individuals.

- They know exactly what is expected (product standards and process standards).
- They know their actual performance (timely feedback).
- They are able to regulate the process because they have:

A capable process.

The necessary materials, tools, skills, and knowledge.

The authority to adjust the process.

[f] Prepare to hold the gains.

4. Design for culture, to minimize or overcome resistance.

[a] Identify likely sources of resistance (barriers) and supports (aids). Resistance typically arises because of:

- Fear of the unknown
- Unwillingness to change customary routines
- The need to acquire new skills
- Unwillingness to adopt a remedy "not invented here"
- Failure to recognize that a problem exists
- Failure of previous solutions
- Expense.

[b] Rate the barriers and aids according to their perceived strengths.

[c] Identify the countermeasures needed to overcome the barriers. Consider:

- Providing participation
- Providing enough time
- Keeping proposals free of excess baggage
- Treating employes with dignity
- Reversing positions to better understand the impact on the culture
- Dealing with resistance seriously and directly

5. Install statistical process control (spc) where necessary, to assure that your process remains stable and predictable, and runs in the most economic manner.

6. Consider introducing 5s standards, to make the workplace function smoothly with maximum value-added activity and minimum nonvalue activity.

Validate measurement system. Utilize commercially available software such as Minitab Attribute Gage R&R Effectiveness (as in Measure phase), to assure that the measurements utilized to evaluate control subjects can be depended on to tell the truth.

Establish process capability

1. Prove effectiveness of the new, improved process, to assure that the new controls work; and to discover if your original problem has improved and assure no new problems have inadvertently been created by your improvement(s).

   [a] Decide how the improvements will be tested.

   • Agree on the type of test(s)

   • Decide when, how long, and who will conduct the test(s)

   • Prepare a test plan for each improvement

   [b] Identify limitations of the test(s).

   [c] Develop an approach to deal with limitations.

   [d] Conduct the test.

   [e] Measure results.

   [f] Adjust the improvements if results are not satisfactory.

   [g] Retest, measure, and adjust until satisfied that the improved process will work under operating conditions.

2. Utilizing control charts, assure that the new process is in statistical control with respect to each individual control subject. If not, improve the process until it is.

3. When, and only when, the process is in statistical control, utilize Minitab Capability Six Pack—as in the Measure

phase—to determine process capability for each individual control subject.

Implement improvements and controls, and monitor

1. Transfer to the operating forces all the updated control plans, etc., and train the people involved in the process in the new procedure.

    [a] Develop a plan for transferring the control plan to the operating forces. The plan for transferring should indicate:

    • How, when, and where the improvements will be implemented

    • Why the changes are necessary and what they will achieve

    • The detailed steps to be followed in the implementation

    [b] Involve those affected by the change in the planning and implementation.

    [c] Coordinate changes with the leadership team, black belt, champion, executive council, and the affected managers.

    [d] Ensure preparations are completed before implementation, including:

    • Written procedures

    • Training

    • Equipment, materials, and supplies

    • Staffing changes

    • Changes in assignments and responsibilities

    [e] Monitor the results.

2. Periodically audit the process, and also the new controls, to assure holding the gains.

    [a] Integrate controls with a balanced scorecard.

    [b] Develop systems for reporting results. When developing systems for reporting results, determine:

- What measures will be reported
- How frequently
- To whom (should be a level of management prepared to monitor progress and respond if gains are not held)

[c] Document the controls. When documenting the controls, indicate:

- The control standard
- Measurements of the process
- Feedback loop responsibilities (who does what if controls are defective)

3. After a suitable period of time, transfer the audit function to the operating forces, and disband the team (with appropriate celebrations and recognition).

Design effective controls

1. Provide the means to measure the results of the new process.
   - End-result measures
   - In-process measures
2. Establish the control standard for each measure. Base each control standard on the actual performance of the new process.
3. Determine how the actual performance will be compared to the standard.
4. Design actions to regulate performance if it does not meet the standard. Use a control spreadsheet to develop an action plan for each control variable.
5. Establish self-control for individuals.
   - They know what is expected
   - They know their actual performance
   - They are able to regulate the process because they have:
     A capable process

The necessary tools, skills, and knowledge

The authority

MAKING DMAIC HAPPEN: ROLES AND TRAINING

Executing a Six Sigma effort requires building a suitable infrastructure. A number of key roles are operating within the infrastructure. Each role is essential yet, by itself, insufficient to produce the improvement an organization expects from DMAIC. Each role requires training. The roles of black belt and master black belt require certification as well. Certification is granted upon completing subject matter training, carrying out a number of significant projects, and passing written and oral examinations. The key roles are:

- Executive management
- Champion
- Black belt
- Green belt
- Orange belt
- Yellow belt
- Project team resources
- Master black belt
- Process owner

The roles of *all* the members of the executive leadership, acting as a team, include:

- *Set goals.* Identify the best opportunities to improve performance and set strategic and annual goals for the organization. Establish accountability for meeting goals.
- *Establish infrastructure.* Establish or revise management systems for selecting and assigning projects, organizational

reporting of project progress, accountability of the various roles, performance appraisal, reward, and recognition.

- *Support projects and monitor progress.* Enable project teams to carry out their project missions. Provide the necessary training, resources, facilities, budgets, time, and most important, management support. Monitor progress of projects and keep them on track.
- *Ask the right questions* at each phase of the Six Sigma project.
- *Deal with critical needed changes.*
- *Receive training* in enough detail to be able to support and evaluate the work of all the other roles.

Please bear in mind the discussion in Chapter 9, "Breakthroughs in Culture," where the point is emphasized that all members of the executive team and managers at all levels should be committed to the Six Sigma effort, agree to support it, and act with unified focus and consistency to facilitate the gradual cultural changes that will inevitably be required. A fractured executive and management team can and usually does wreak havoc and confusion on a Six Sigma effort, drains the energy out of those trying to make it succeed, and leaves in its wake disillusionment and meager results. The executive team loses its credibility and ability to lead (assuming it ever had any credibility).

Champions are usually members of management (or at least folks with organizational clout). The champion:

- Identifies improvement projects that meet strategic goals
- Is responsible for project problem statements and mission statements
- Identifies and selects competent black belts and team members
- Mentors and advises on prioritizing, planning, and launching Six Sigma projects
- Mentors and coaches black belts
- Removes organizational obstacles that may impede the work of the black belts or the project teams

- Provides approval and support to implement improvements designed by the project teams
- Provides recognition and rewards to the black belts and teams upon successful completion of their projects
- Communicates with executive management and peers the progress and results associated with the Six Sigma efforts
- Knocks down barriers the teams encounter
- Understands and upholds the Six Sigma methodology
- In general: manages, supports, defends, protects, fights for, maintains, upholds, and advocates for Six Sigma efforts

Usually, a strong champion is found behind successful projects. Weak champions are usually associated with weak results. Effective champions are good leaders (see Chapter 6, "Breakthroughs in Leadership").

Black belts are on-site implementation experts with the ability to develop, coach, and lead cross-functional process improvement teams. They mentor and advise management on Six Sigma issues. Black belts have an in-depth understanding of Six Sigma philosophy, theory, strategy, tactics, and Six Sigma tools. The training to be certified as a black belt is rigorous and demanding. An illustrative list of topics would include:

- Project-critical team leadership and facilitation skills
- Six Sigma methodology
- Quality improvement tools
- Use of an appropriate statistical software package
- Measurement system analysis
- Determining process capability
- Process mapping
- Quality function deployment
- Failure mode, effect, and criticality analysis
- Correlation and regression
- Hypothesis testing using attribute and variables data

- Analysis of variance
- Design of experiments
- Evolutionary operations
- Quality systems
- Lean enterprise
- Dominant variables
- Mistake-proofing
- Statistical process control
- Process control plans
- Self-control

Armed with this training—usually delivered in four week-long sessions with four- to five-week intervening intervals devoted to carrying out a real Six Sigma project—the roles of the black belt include:

- Leading and facilitating project teams
- Leading projects
- Training teams in Six Sigma methodology and use of tools
- Using Six Sigma methodology to solve problems
- Coaching team members and others as required
- Performing project management
- Continuously communicating with champions and impacted managers
- Networking with other black belts

A master black belt receives training and coaching beyond that of a black belt. Master black belts are qualified to train black belts. The roles of master black belt include:

- Acting as internal Six Sigma consultant, trainer, and expert on Six Sigma
- Managing and facilitating multiple projects—and their black belts

- Supporting and advising champions and executive management

- Providing technical support and mentoring as needed

Everyone else in the organization—those who are not champions, master black belts, or black belts—become either a green belt, orange belt, yellow belt, or white belt. Space precludes delineating distinctions here. Suffice it to say that the different colored belts vary according to the amount of formal training received and the active roles each takes in participating in Six Sigma activities. In an ideal situation, all organization members receive training at some minimal level, and are awarded the appropriate belt. Everyone feels included, and everyone understands what Six Sigma is all about, and just as important, what it is not about. No one is left out, to wonder what Six Sigma is all about, or to resent or resist it. This unifies the organization behind the Six Sigma effort and significantly reduces pockets of resistance.

HIGH POINTS OF "BREAKTHROUGHS IN CURRENT PERFORMANCE"

- Breakthroughs in current performance solves quality problems: excessive number of defects, excessive delays, excessively long time cycles, and excessive costs.
- The route to breakthroughs in current performance requires two journeys: the diagnostic journey and the remedial journey. These "journeys" represent the application of the scientific method to the solution of performance problems.
- The diagnostic journey proceeds as follows:

From problem to symptoms of the problem

From symptoms to theories of causes of the symptoms

From theories to testing of the theories

From tests to establishing root cause(s) of the symptoms

- The remedial journey proceeds as follows:

  From root cause(s) to design of remedies of the cause(s)

  From design of remedies to testing and proving the remedies under operating conditions

  From workable remedies to dealing with predictable resistance to change

  From dealing with resistance to establishing new controls on the remedies to hold the gains

- The classic "Juran" model of performance (quality) improvement follows these basic steps:

  Identify a project. (Management does this.)

  Establish a project. (Management does this.)

  Diagnose the cause(s). (Project team does this.)

  Remedy the cause(s). (Project team plus work group where cause[s] originate do this.)

  Hold the gains. (Project team and effected operating forces do this.)

  Replicate results and nominate new projects. (Management does this.)

- The Six Sigma model of performance (quality) improvement follows these basic steps:

  Define (Establish a project). (Champions and management do this.)

  Measure (Problem to symptoms). (Project team does this.)

  Analyze (Discover root cause[s]). (Project team does this.)

  Improve (Remedy the cause[s]). (Project team and others do this.)

  Control (Hold the gains). (Project team and others do this.)

- In both models, the basic roles of management are:

  Establish quality and executive councils

  Select projects

  Write problem and mission statements

Provide resources, especially time, to carry out the project

Assign teams, team leaders, and facilitators/black belts and projects to teams

Review progress and remove barriers and resistance

Provide recognition and rewards

- In both models, the basic roles of project teams are:

Analyze symptoms

Theorize as to causes

Test theories

Establish root cause(s)

Stimulate remedies and controls

- Project selection requires some know-how on the part of management so "doable" projects are identified, and the team clearly understands both the problem and the mission.

- In the ideal, all members of an organization engaged in a Six Sigma effort will receive at least minimal training or orientation, and subsequently be awarded the appropriate belt. In this way, everyone understands the Six Sigma effort; no one feels left out or in the dark. Consequently, the organization becomes unified behind the Six Sigma effort and resistance is minimized.

BREAKTHROUGHS
IN CULTURE

Breakthroughs in culture address the basic question: "How do I create a social climate that encourages organization members to eagerly march together toward the organization's performance goals?"

An organization's culture exerts an extraordinarily powerful impact on organizational performance. The culture determines what is right or wrong, legitimate or illegitimate, and acceptable or unacceptable. Consequently, breakthrough in culture is profoundly influential in achieving performance breakthrough. It is also probably the most difficult and time-consuming type of breakthrough to make happen. And, it is so widely misunderstood that attempts to pull it off often fail.

Breakthrough in culture: a) creates a set of behavior standards, and a social climate that supports organizational goals; b) instills in all functions and levels the values and beliefs that guide organizational behavior and decision-making; and c) determines organizational cultural patterns such as *style* (informal versus formal, flexible versus rigid, authoritarian top-down versus participative collaboration, management-driven versus leadership-driven, and the like), the organization's *caste system* (the relative status of each function), and the *reward structure* (who is rewarded for doing what).

CULTURE DEFINED

Your organization is a society. A society is "an enduring and cooperating social group whose members have developed organized patterns of relationships through interaction with each other"... a group of people engaged in a common purpose."

Your workplace is a society, and as such, it is held together by the shared *beliefs* and *values* that are deeply embedded in the personalities of the society's members.

(A workplace whose workforce is segmented into individuals or groups who embody conflicting beliefs and values doesn't hold together. Various social explosions will eventually occur: resistance, revolts, mutinies, strikes, resignations, transfers, firings, divestiture, bankruptcy, etc.).

Society members are rewarded for conforming to their society's beliefs and values—its norms—and they are punished for departing from them. Norms encompass not only values and beliefs, but they include enduring systems of relationships, status, customs, rituals, and practices.

Societal norms are so strong and deeply embedded that they lead to customary patterns of social behavior sometimes called "cultural patterns." In the workplace, one can identify performance-determining cultural patterns such as: participative versus authoritarian management styles; casual versus formal dress and conversational styles ("Mr./Ms." and "Sir/Madam" versus first names); high trust level that makes it safe to say what you really think versus low trust level/suspiciousness that restricts honest or complete communication and breeds game-playing, deceit, and confusion.

WHAT DOES CULTURE HAVE TO DO WITH MANAGING AN ORGANIZATION?

To achieve performance breakthrough, it is desirable—if not necessary—that the organization's norms and cultural patterns support the organization's performance goals. Without this support, performance goals may well be diluted, resisted, indifferently pursued, or simply ignored. For these reasons, the characteristics of your organization's culture is a vital matter that your management needs to understand and be prepared to influence. As we shall see, this is easier said than done; but it *can* be done.

A timely example of the influence of culture on an organization's performance is provided by J. M. Juran. Here are excerpts from his description of a management challenge currently facing managers, as it has for many years: getting acceptance on the shop floor for statistical control charts, typically a key element in the Control phase of Six Sigma.

(Control charts detect the pattern of variation exhibited by a repetitive process. They can provide a great deal of information about the performance of a process—information unobtainable from any other source. They are widely used in manufacturing and in all kinds of repetitive transactional processes such as those found in hospitals and offices. Among other things, control charts inform the employee if and when to adjust the process, a feature which largely replaces the traditional practice of the employee making this decision. On top of that, control charts are based on the laws of probability and statistics, topics that are widely misunderstood or regarded as impenetrable mysteries.)

"There has been great difficulty in getting production operators and supervisors to accept control charts as a shop tool. I believe this to be a statement of fact, based on extensive first-hand observation of the shockingly high mortality rate of control charts when actually introduced on the shop floor.

"This difficulty is not merely a current phenomenon. We encountered it back in the late 1920s in the pioneering effort to use control charts on the production floor of the Hawthorne Works of the Western Electric Company. Neither is it merely an American phenomenon, since I have witnessed the same difficulty in Western Europe and in Japan as well....

"It is my belief that the failure of the control chart to secure wide acceptance on the factory floor is due mainly to lack of adaptation into the culture of the factory, rather than to technical weaknesses in the control chart....

"There are a number of problems created by the control chart, as viewed by the shop supervisor:

- The control chart lacks "legitimacy" (i.e., it is issued by a department not recognized as having industrial legislative powers).
- The control chart conflicts with the specification, leaving the operator to resolve the conflict.
- The control chart is in conflict with other forms of data collection and presentation, leaving the operator to resolve the conflict.
- The control chart calls for a pattern of operator action that differs from past practice, but without solving the new problems created as a result of disturbing this past practice.

Legitimacy of the control chart. The human passion for "law and order" does not stop at the company gate. Within the plant, there is the same human need for a predictable life, free from unpleasant surprises. Applied to the production operator, this concept of law and order resolves into various principles:

1. There must be one and only one personal supervisor (boss) to whom he/she is responsible.
2. There is no limit to the number of impersonal bosses (manuals, drawings, routines), but each of these must be legitimate, that is, it must have clear official status.
3. When there is a conflict between the orders of the personal boss and an impersonal boss, the former prevails.
4. When there is a conflict between something "legitimate" and something not established as legitimate, the former prevails.

"There can be no quarrel with these principles, since they are vital to law and order on the factory floor....

"...Introduction of control charts to the factory floor results in a series of changes in the cultural pattern of the shop:

- A new source of industrial law is opened up, without clear evidence of its legitimacy.

- This new industrial law conflicts with long-standing laws for which there has been no clear repeal through recognized channels of law.
- New sources of factual information are introduced, without clear disposition of old sources.
- New duties are created without clear knowledge of their effect on those who are to perform those duties....

Conclusions. Introduction of modern quality control techniques has an impact on the factory in two aspects:

- The technical aspect, involving changes in processes, instrument records, and other technical features of the operation.
- The social aspect, involving changes in humans, status, habits, relationships, scale of values, language, and other features of the cultural pattern of the shop.

"The main resistance to change is due to the disturbance of the cultural pattern of the shop."   —J. M. Juran

HOW ARE NORMS ACQUIRED?

New members of a society—a baby born into a family or a new employee hired into the workplace—are carefully taught who is who and what is what. In short, they are taught the norms and cultural patterns of that particular society. In time, they discover that compliance with the norms and cultural patterns can be satisfying and rewarding. Resistance or violation of the norms and cultural patterns can be very dissatisfying because it brings on disapproval, condemnation, and possibly punishment.

If an individual receives a relatively consistent pattern of rewards and punishments over time, the beliefs and behaviors being rewarded gradually become a part of that individual's personal set of norms, values, and beliefs. Those behaviors that are consistently disapproved or punished will gradually be discarded and not repeated. The individual will have become socialized.

HOW ARE NORMS CHANGED?

Note that socialization can take several years to take hold. This is an important prerequisite for successfully changing an organization's culture that must be understood and anticipated by agents of change, such as top management. The old patterns must be extinguished and replaced by the new ones. This takes time, and consistent, persistent effort. Such are the realities.

Consider what the anthropologist, Margaret Meade, has to say about learning new behaviors and beliefs:

"An effective way to encourage the learning of new behaviors and attitudes is by consistent prompt attachment of some form of satisfaction to them. This may take the form of consistent praise, approval, privilege, improved social status, strengthened integration with one's group, or material reward. It is particularly important when the desired change is such that the advantages are slow to materialize—for example, it takes months or even years to appreciate a change in nutrition, or to register the effect of a new way of planting seedlings in the increased yield of an orchard. Here the gap between the new behavior and results, which will not reinforce the behavior until they are fully appreciated, has to be filled in other ways."

She continues:

"The learning of new behaviors and attitudes can be achieved by the learner's living through a long series of situations in which the new behavior is made highly satisfying—without exception if possible—and the old not satisfying."

And:

"New information psychologically available to an individual, but contrary to his customary behavior, beliefs, and attitudes, may not even be perceived. Even if he is actually forced to recognize its existence, it may be rationalized away, or almost immediately forgotten."

Finally:

"...as an individual's behavior, beliefs, and attitudes are shared with members of his cultural group, it may be necessary to effect a change in the goals or systems of behavior of the whole group before any given individual's behavior will change in some particular respect. This is particularly likely to be so if the need of the individual for group acceptance is very great—either because of his own psychological make-up or because of his position in society."

Implications for achieving breakthroughs in culture are:

- To be most effective, the entire management team, at all levels must share, exhibit, and reinforce desired new cultural norms and patterns of behavior—and they must be consistent, uninterrupted, and persistent.
- Don't expect cultural norms or behavior to change simply because you publish the organization's stated values in official printed material, or describe them in speeches or exhortations. The actual cultural norms and patterns may bear no resemblance at all to the values described to the public or proclaimed in exhortations. The same is true of the actual flow of influence compared to the flow shown on the organization chart. (New employees rapidly learn who is really who and what is really what, in contrast to and in spite of the official publicity.)

A forceful leader-manager can, by virtue of his/her personality and commitment, influence the behavior of individual followers in the *short term* with rewards, recognition, and selective exclusion from rewards. The authors know of organizations who, in order to introduce a Six Sigma or similar effort, have presented messages to their employees along the following lines:

"The company cannot tell you what to believe, and we are not asking you to believe in our new Six Sigma initiative, although we hope you do. We can, however, expect you to behave in certain ways with respect to it. Therefore, let it be known that you are expected to support it, or at least get

out of its way, and not resist. Henceforth, rewards and promotions will go to those who energetically support and participate in the Six Sigma activities. Those who do not
support it and participate in it will not be eligible for raises
or promotions. They will be left behind, and perhaps even
replaced with others who do support it."

This is fairly strong language. Such companies often achieve some results in the short term. However, should a forceful leader depart without causing the new initiative to become embedded in the organization's cultural norms and patterns (to the extent that individual members have taken on these new values and practices as their own), it is not unusual for the new thrust to die out for lack of consistent and persistent reinforcement.

RESISTANCE TO CHANGE

Curiously, even with such reinforcement, change, even beneficial change, will often be resisted. The would-be agent of change needs to understand the nature of this resistance and how to prevent or overcome it.

The control chart case example at the beginning of this chapter drew the conclusion that the main resistance to change is due to the disturbance of the cultural pattern of the shop when a change is proposed or attempted. People who are successful—and therefore comfortable—functioning in the current social or technical system don't want to have their comfortable existence disrupted, especially by an "illegitimate" change.

When a technical or social change is introduced into a group, group members immediately worry that their secure status and comfort level under the new system may be very different (worse) than under the current system. Threatened with the frightening possibility of losing the ability to perform well or losing status, the natural impulse is to resist the change. They have too much at stake in the current system. The new system will require them not only to let go of the current system willingly, but also to embrace the uncertain, unpredictable new way of performing. This is a tall order.

It is remarkable how profoundly even a tiny departure from cultural norms will upset society members. An acquaintance of one of the authors, a Professor of Sociology, asked students of his introductory class to go into the community and violate the smallest social norm they could think of: nothing immoral, illegal, or obnoxious, just a seemingly inconsequential, insignificant norm. One student knocked on the door of a residence neighboring the campus and asked to borrow a toothbrush. Another appeared at someone else's front door and asked the perfect stranger who answered the doorbell to lend him his pipe so he could have a smoke. Both local residents, long familiar with the eccentricities of college students, nevertheless were so frightened and upset by this "weird" behavior that they called the police, who in turn complained to the Dean of the College and asked him to restrain Professor X.

Participants in one of the author's seminars experience firsthand the extraordinary power of cultural resistance to a very small change. An exercise during a section of the course devoted to the topic "Resistance to Change" occurs on the fourth day of a five-day course.

Invariably, course participants take the same place around the seminar table each day. By the fourth day, this place has become their space, and is protected from squatters with territorial zeal.

The instructor, with no expression on his/her face, introduces the session with an order for student A to exchange places with student B, across the table. Student C is ordered to change places with student D, and so on. No introduction to prepare the people for this move is given. No explanation is provided. No apologies for the sudden inconvenience are made. Just orders to move, *now*.

A scribe records the behavior of those ordered to move. Typical observed behaviors include:

- Nervous laughter
- Muttering under the breath

- Attempts at weak jokes, seemingly to break tension
- Wisecracks
- People, sometimes begrudgingly judging from their facial expressions, move

Now a second instructor intervenes, reminding the first instructor: "What you did when having the people move was not what we discussed." The second instructor attempts to rearrange the rearrangement by ordering each person to move a second time. Now the intensity and diversity of observed behaviors increases. In addition to the behaviors already mentioned, new ones emerge:

- Loud complaints
- Loud vigorous objections
- Profanity
- Questions: "What is going on?"
- Challenges: "Why are we doing this?"
- Denunciation: "This is silly!"
- Resistance: one or two people simply refuse to move, loudly and defiantly

After (very) politely offering the suggestion that the participants may wish to return to their personal seat, the instructor poses a question: "Why were you so upset, when all you were asked to do was move 10 feet?"

The answers pour out:

- "I didn't know why we were doing this."
- "I was a little fearful because I didn't know what was going to happen next."
- "I suddenly didn't know what the rules were, any more."
- "You told me what to do. You didn't ask. I don't like being arbitrarily ordered to do something."

- "I was insecure, because I was not in control, and confused."
- "It was too much, outrageous."

At this point, the two instructors ask: "What could we have done different or differently to get you to move without being so upset?" The typical responses are instructive:

- "You could have explained the need for us to move."
- "You could have asked us for suggestions of what to do or how to do it, once we understood the need."
- "You could have asked, instead of arbitrarily ordering us. I didn't even feel like a person when you just told me what to do. It's been a long time since someone just told me to do something. It was impolite, insulting, demeaning, and it was wrong!"
- "You could have given us a little time to get used to the idea of having to move. It was all so sudden and unexpected."

WHAT DOES RESISTANCE TO CHANGE LOOK LIKE?

Some resistance is intense, dramatic, even violent. Juran reminds us of some examples: When fourteenth-century European astronomers postulated a sun-centered universe, this idea flew in the face of the prevailing cultural beliefs in an Earth-centered universe. This belief had been passed down for many generations by their ancestors, religious leaders, grandparents, and parents. (Furthermore, on clear days, one could see with one's own eyes the sun moving around the Earth.) Reaction to the new "preposterous" unacceptable idea was swift and violent. If the sun-centered believers are correct, then the Earth-centered believers are incorrect—an unacceptable, illegitimate, wrong-headed notion. To believe in the new idea required rejecting and tossing out the old. But the old was deeply embedded in the culture. So the blasphemous astronomers were burned at the stake.

Another example from Juran: When railroads converted from steam-powered to diesel-powered locomotives in the 1940s, railroad workers in the United States objected. It is unsafe, even immoral, they protested, to trust an entire train-load of people or valuable goods to the lone operator required to drive a diesel. Locomotives had "always" been operated by two people, an engineer who drove, and a fireman who stoked the fire. If one *were* incapacitated, the other could take over. But what if the diesel engineer had a heart attack and died? So intense were the resulting strikes that an agreement was finally hammered out to keep the fireman on the job in the diesels!

Of course the railroad workers were really protesting the likely loss of their status and jobs.

In an organization, resistance to change can be open and direct or subtle, but the result is the same—to delay, disrupt, or prevent a proposed change in the cultural pattern. Here are some of the more obvious direct forms of organizational resistance to change:

- Bitter confrontation and arguments in meetings
- Resignations
- Transfers
- Strikes or other job actions
- Firings
- Whistle-blowing, particularly to the press

More subtle forms of resistance to change include:

- Failing to return a phone call or answer an e-mail
- Failing to keep appointments
- Assigning to a subordinate a task that prevents complying with a change (such as loading up a staff person with work so he/she is too busy to attend Six Sigma project team meetings)
- Being late to meetings
- Ignoring meetings altogether; not showing up

- Promising something (such as sales figures or data) and not delivering
- Ignoring the change (such as parking in the same old close-by place instead of the newly assigned far-off place, wearing the same casual clothes instead of newly required formal clothes, keeping the same old hours instead of the newly established hours, and failing to produce a newly requested weekly report)
- Writing expense reports in long hand instead of using the new computer software package

PREVENTING OR OVERCOMING RESISTANCE TO CHANGE

Juran Institute employs the "Rules of the Road" for handling resistance to change.

- Secure the active participation of those who will be affected, both in the planning and in the execution of the change. Consider also including third parties who can supply perspective, balance, and objectivity.
- Provide sufficient time for the mental changes required for coming to terms with the idea of a change.
- Start small and proceed gradually, allowing people to experience success and satisfaction from small increments of change.
- No surprises; don't suddenly dump an unexpected change on anyone. This is almost guaranteed to provoke resistance.
- Choose the right time (no painful cost-cutting when executives are getting big bonuses and don't feel any pain themselves, for example).
- Strip off all technical cultural baggage not strictly needed for introducing the change. If possible, refrain from changing several things at once. Many simultaneous changes are confusing and difficult to handle, and could trigger resistance.

- Work with the recognized leadership of the culture (who may be different from the organization chart). Deal with "who is really who."
- Treat the people with dignity. They usually respond in kind.
- Put yourself in the other person's place. You may even engage in some role-playing to "get into their head."

NORMS HELPFUL IN ACHIEVING PERFORMANCE BREAKTHROUGH

Achieving performance breakthrough requires a highly supportive culture. Certain cultural norms appear to be instrumental in providing the needed support. If these norms are not now part of your culture, some breakthroughs in culture may be required to implant them. Here are some of the more enabling norms:

- A *belief that the quality of a product or process is at least of equal importance, and probably of greater importance than the mere quantity produced.* This belief results in decisions favoring quality: defective items do not get passed on down the line or out the door; chronic errors and delays are corrected, etc.
- A *fanatic commitment to meeting customer needs.* Everyone knows who his/her customers are (those who receive the results of their work), and how well he/she is doing at meeting those needs (They *ask.*). Organization members, if necessary, drop everything and go out of their way to assist customers in need.
- A *fanatic commitment to stretch goals and continuous improvement.* There is always an economic opportunity for improving products or processes. Those who practice continuous improvement keep up with, or become better than competitors. (Those organizations that do not practice continuous improvement fall behind and become irrelevant or worse.) Six Sigma product design and process improve-

ment is capable, if executed properly, of producing superb economical designs, and nearly defect-free processes to produce them. Result: Very satisfied customers and sharply reduced costs. The resulting sales and savings show up directly on the bottom line.

- *A belief that there should be no "sacred cows."* Everything that occurs in an organization is subject to challenge and revision, if it no longer makes sense under new realities. That includes even products or practices initiated by important organization members. It even includes the core business or mission of the organization, should that have become irrelevant or obsolete in the face of societal or technological change.

- *A customer-oriented code of conduct and code of ethics.* This code is published, taught in new employee orientations, and taken into consideration in performance ratings and in distributing rewards. Everyone is expected at all times to behave and make decisions in accordance with the code. It is enforced, if needed, by managers at all levels. The code applies to everyone, even board members, perhaps especially to them considering their power to influence everyone else.

- *A belief that continuous adaptive change is not only good, but necessary.* To keep alive, you must develop a system for discovering social, governmental, international, or technological trends that could have an impact on your organization. In addition, you will need to create and maintain structures and processes that enable quick effective response to these newly discovered trends.

Given the difficulty of predicting trends in the fast-moving contemporary world, it becomes vital for organizations to have such processes and structures in place and operating. If you fail to learn and appropriately adapt to what you learn, your organization can be left behind very suddenly and unexpectedly, and end up in the scrap heap. Many former e-commerce companies come to mind. Many rusting abandoned factories— the world over—testify to the consequences of not keeping up and consequently being left behind.

Adaptive activities and structures are described in Chapter 10, "Breakthroughs in Adaptability."

CULTURAL PATTERNS HELPFUL IN ACHIEVING PERFORMANCE BREAKTHROUGH

A number of cultural patterns have been identified by experience and research as influential contributors to reaching the excellent levels of performance embodied in performance breakthrough. If these patterns of behavior are not now characteristic of your organization, you may wish to consider achieving breakthroughs in culture to "convert" your organization to these modes of operation. Enabling cultural patterns include:

- *A collaborative, as opposed to a competitive mode of performing work.* Many jobs are now described by team job descriptions, rather by individual job descriptions. Defined are the goals and the tasks required of teams, not the individuals who constitute the team. The performance produced by the synergy of diverse contributing team members can far exceed the performance that the same individuals could produce, in aggregate, operating by themselves.

- *A generally participative, as opposed to a generally authoritarian management style.* Authoritarian management reserves decision making to managers, not doers, not even leaders. It breeds passivity, indifference, and dependency of workers on their managers. Decisions can be delayed, as requests for answers from subordinates accumulate on bosses' desks. While waiting for direction from above, productivity lags, problems do not get addressed, and adaptability is severely restricted. Initiative and enthusiasm for the job is stifled, possibly extinguished, especially if doers are scolded, punished, or overridden when they put forth personal opinions or make independent decisions. Leaders are expressly discouraged and blocked from exerting leadership. Communication as well as authority tends to be vertical, largely top-down, and relatively uninformed, because the feared authoritarians don't get told everything they need to know. One can say that the more authoritarian an organization is, the less likely the managers are playing with a full deck (of information). It's too risky sometimes for subordinates to give it to them. Scientific and business literature contains numerous descriptions of debilitating organizational pathologies that are created by fear of saying what you want to say (what you really think, what you feel).

Participative management, by and large, resembles essentially the opposite of authoritarian management. Authoritarians will be pleased to know that those at the top of the authority pyramid in a participative organization don't necessarily lose any power relative to others. In fact, because of the wider communication that takes place in all directions, they gain power, because they have more and better information upon which to base decisions. And the decisions themselves are often shared with others closer to the issue being decided. Ironically, the total amount of power exercised in a participative organization is greater than the total amount of power exercised in an authoritarian organization (see Figure 9.1).

Note that as the total amount of power expands, the relative power of manager X under both systems remains the same.

- *A high level of trust (feeling safe), as opposed to a high level of fear (feeling unsafe, unwilling to offer true opinions, or take stands or risks).* To achieve this cultural pattern, messengers bearing "bad" news must not be shot, especially by the top manager, or anyone else in the organization, for that matter. Obviously, to punish messengers is a very effective means of shutting off communication. To avoid this, in order to feel safe, people must be rewarded for telling the truth, venturing their personal opinion, delivering bad news, or taking initiative. Utterances or actions can be constructively criticized (so as to improve upon them), but never destructively criticized (public put-downs, mockery, accusations of stupidity or ignorance, etc.).

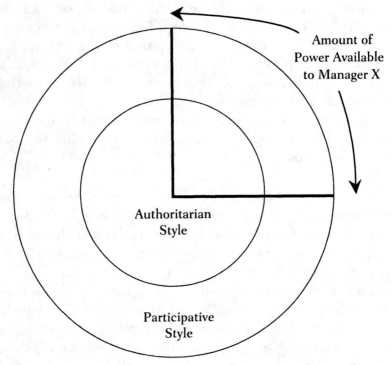

FIGURE 9.1   Total amount of power exercised in an organization.

QUALITY POLICY AND CULTURAL NORMS

Policies are guides for decision making. Quality system manuals typically begin with a statement of the organization's quality policy. This statement rates the relative worth that organization members should place on producing high-quality products, as distinguished from the mere quantity of products produced. ("High-quality products" are goods, services, or information that meet important customer needs at the lowest optimum cost with few if any defects, delays, errors, etc.). High-quality products produce customer satisfaction, sales revenue, repeat demand or sales, and low costs of poor quality (unnecessary waste). Here, in that one sentence, are the reasons for attempting quality improvement.

The inclusion of the value statement in your organization's quality manual reinforces some of the instrumental cultural norms and patterns essential for achieving a "quality culture" and ultimately performance breakthrough.

Keep in mind that if the value statement, designed to be a guide for decision making, is ignored and not enforced, it becomes worthless, except perhaps as a means of deceiving customers and employees in the short term. But you can be sure that customers and employees will soon catch on to the truth. They will dismiss the quality policy, waving it away as a sham that diminishes the whole organization and degrades the credibility of the management.

HUMAN RESOURCES ADMINISTRATION AND CULTURAL PATTERNS

The human resources function plays a significant role in reinforcing cultural norms. It does so by several means that include:

- *Recruiting.* Advertisements contain descriptions of desirable traits (e.g., dependable, energetic, self-starter, creative, analytic, etc.), as well as characterizations of the organization (e.g., service-oriented, customer-oriented, committed to being a world leader in quality, progressive, world-class, equal opportunity, etc.). Organizational values are often featured in these messages.

- *Orientation and training.* It is customary when providing new employees with an introduction to an organization to review with them expected modes of dress, behavior, attitudes, traditional styles of working together, etc.

- *Publishing employee handbooks.* The handbooks distributed to new employees, and to everyone annually, are replete with descriptions of organizational history, traditional policies and practices, and expectations for organization members. All of these topics express directly or indirectly detailed elements of the official culture.

- *Publishing newsletters.* The choice of events that constitutes news is a pretty clear indication of what an organization considers important and worthwhile, officially at least. Think of typical content: anniversaries of service, recognition for long service, news about employees' families, news of promotions, news of the accomplishments of project teams or particular work groups, special activities such as employee sports teams, hobbies or community service, etc. If you reflect upon the nature of the events that are reported, it can tell you something about the culture. Indeed, newsletters are designed to promote organizational values such as coherence and a sense of "family."

- *Reward and recognition practices.* In our rapidly changing world, management teams find themselves agonizing over what kind of employee behavior should be rewarded. Whatever the behavior is, when it is rewarded, the reward reinforces the cultural norms embodied in that behavior, and it should induce more of the same behavior from the rewarded ones, as well as attract others to do the same.

- *Career path and promotion practices.* If you track the record of those promoted in an organization, you are likely to find either [a.] behavior that conforms to the traditional cultural norms in their background, or [b.] behavior that resembles desired new cultural norms required for a given organizational change, such as launching a Six Sigma effort. In the former case, management wants to preserve the current culture; in the latter case, management wants to create breakthroughs in culture and bring about a new culture that is at least somewhat altered. In both cases, the issue of the relationship of the person being promoted to the organizational culture is a significant factor in granting the promotion.

HIGH POINTS OF "BREAKTHROUGHS IN CULTURE"

- Your organization is a society. Societies are groups of people engaged in a common purpose and held together by its culture: shared beliefs, practices, customs, traditions, norms, statuses, and values (ideas of what is right and wrong, legitimate or illegitimate, etc.).

- An organization's culture exerts a powerful, even a determining impact on organizational performance.

- Your organization's culture must support organizational performance goals. If it doesn't, official performance goals, and performance itself, will be diluted, resisted, indifferently pursued, or simply ignored.

- A workplace whose work force is segmented into individuals or groups who embody conflicting norms, beliefs, and values doesn't hold together. Various costly social explosions eventually occur: resistance, revolts, mutinies, strikes, resignations, transfers, firings, divestiture, bankruptcy, etc.

- Cultural norms are inculcated into children and into new employees by receiving, over time, a consistent pattern of rewards for behavior that conforms to current norms, and punishments for behavior that does not conform.

- To change an adult's norms or conformance to a new cultural pattern requires not only consistent rewards for conforming to the new norms, but also consistent punishment (or withholding of rewards) for conforming to the old norms.

- To be most effective, the entire management team, at all levels must share, exhibit, and reinforce desired new cultural norms and patterns of behavior—and they must do this consistently and persistently over considerable time periods (months, years).

- Management teams cannot change their organization's culture simply by publishing new "official" norms and cultural behavior and exhorting everyone to follow them.

- If change, even beneficial change, is imposed upon an organization, it is likely to be resisted. Reasons: fear of the unknown, resentment at not being prepared for or involved in the change, and judging that the change violates current norms and is therefore illegitimate or wrong.

- Guidelines for managers to prevent or overcome resistance to change:

Provide participation

Provide sufficient time

Start small and expand purview of the change

No surprises

- Choose the right time.

No excess baggage; keep the change simple

Work with the recognized leadership, formal or informal

Treat the people with dignity

Try putting yourself in the other person's place

- Norms helpful for achieving performance breakthrough include:

    A belief that the quality of a product or process is at least of equal importance, and probably of greater importance than the mere quantity produced

    A fanatic commitment to meeting customer needs

    A fanatic commitment to stretch goals and continuous improvement

    A belief that there should be no "sacred cows"

    A customer-oriented code of conduct and code of ethics

    A belief that continuous adaptive change is not only good, but necessary

- Cultural patterns helpful in achieving performance breakthrough include:

    A collaborative, as opposed to a competitive mode of performing work

    A generally participative, as opposed to a generally authoritarian management style

    A high level of trust as opposed to a high level of fear

- Human resources policies and practices, from recruitment to retirement, must clearly explain and embody the cultural norms and behavior patterns expected by the organization.

BREAKTHROUGHS IN ADAPTABILITY

The survival of your organization, like all open systems, depends on its ability to detect and react to threats and opportunities that present themselves from within and from outside. To detect potential threats and opportunities, an organization must not only gather data and information about what's happening, but it must also discover the (often) elusive meaning and significance the data hold for the organization. Finally, it must take appropriate action to minimize the threats and exploit the opportunities gleaned from the data and information. To do all this will require appropriate organizational structures, some of which may already exist (an intelligence function, an adaptive cycle, a Data Quality Council) and a data quality system. The Data Quality Council acts, among other things, as a "voice of the market" council.

Data are defined as "facts" (such as name, address, age) or "measurements of some physical reality, expressed in numbers and units of measure." These measurements are the raw material of information, which is defined as "answers to questions," or the "meaning revealed by the data, when analyzed."

The typical contemporary organization appears to the authors to be awash in data but bereft of useful information. Even when in possession of multiple databases, much doubt exists regarding the quality of the data and therefore its ability to tell the truth about the question it is supposed to answer.

Plant managers dispute the reliability of production reports, quality reports, especially if the messages contained in the data are unfavorable. Department heads question the accuracy of financial statements, and sales figures, especially when they bring bad tidings.

Often, multiple databases will convey incongruent or contradictory answers to the same question, because each individual database has been designed to answer questions couched in the unique dialect, or based on the unique definitions of terms used by one particular department or function, but not all functions. Data often are stored (hoarded?) in isolated unpublicized pockets, out of sight of the very people in other functions who could benefit from them if they knew they existed.

Anyone who relies on data for making strategic or operational decisions is rendered almost helpless if the data are not available, or are untrustworthy. How can a physician decide on a treatment if X-rays and test results are not available? How can the sales team plan promotions when it does not know how its products are selling compared to the competition? What if these same sales people knew that the very database that could answer their particular questions already exists, but is used for the exclusive benefit of another part of the organization?

It is clear that making breakthroughs in adaptability is difficult, if one cannot get necessary data and information, or if one cannot trust the truthfulness of the information one does get. Some organizations for which up-to-date and trustworthy data are absolutely critical go to great lengths to get useful information. And yet, in spite of their considerable efforts, many nevertheless remain plagued by chronic data quality problems.

Consider the case of an enormous multinational producer of mass-marketed consumer goods. For confidentiality reasons, we will call this organization "Braggabit." Braggabit employed dozens of highly skilled database managers and staff who were organized into several administrative units in several functions. Its common mission was providing upper management with up-to-date information on the behavior of the market. It produced intelligence mainly on: 1) the behavior of ultimate users (their brand preferences and purchases); 2) the volume of like product produced and shipped by themselves and their competitors; 3) on the volume of product, by brand, that was received and shipped by distributors; and 4) the volume of product, by brand, received and sold by retailers.

This information was then—on a weekly basis—assembled, analyzed, and published in standardized summary reports, which Braggabit's upper management utilized weekly to determine strategic and operational plans such as *pricing* by product; *promotions* by product and geographic location or demographic segment; *advertising* campaigns (Braggabit did like to brag a bit in its advertising); and the like.

The quantity of data Braggabit required was so large that it was forced, for cost reasons, to outsource the creation of the data. Subcontracted suppliers weekly gathered and processed data on thousands of retail outlets and millions of users. Separate contractors gathered the raw data, other contractors preprocessed it, and Braggabit stored it, analyzed it, and converted it to information by interpreting it for meaning. In spite of this enormous input of data, and the army of people they employed to gather, process, inspect, and analyze it, over time it became clear that the data as delivered by their suppliers routinely contained many errors of diverse types. Not only were there numerous errors, but some databases, supposedly designed to answer similar questions, came up with numbers that differed from each other, sometimes substantially. No one could tell which database to believe, or if any of them could be believed. These errors in the data from the outside suppliers were so chronic and so numerous that Braggabit came to distrust the data. It had to assign many of its highly paid data experts to inspect and correct the errors, in order to produce a trustworthy summary report that met the upper managers' needs. Time was crucial. Data had to be gathered, processed, analyzed, and published on a weekly basis. All the inspection and correction greatly inflated the costs of the data, and even worse, slowed the process down so much that the upper managers weren't able to have at their disposal data that was current enough for their purposes.

Bad decisions were made on pricing and promotions. Even a small amount of sampling error or nonsampling error could ultimately result in many millions of dollars of unnecessary expenditures, loss of revenue, or both. Sometimes the reports were so late that no decisions could be made, with the same debilitating results. The situation became so perilous and acute that one of the authors can relate this tale to you.

To summarize: This enormous organization, in order to make itself *adaptable* and capable of making rapid adjustments in its vital marketing and sales efforts, spent millions to get up-to-date data on the business environment on a weekly basis. It employed the best data suppliers available and retained the most skilled database experts. In spite of all the good intentions and investment, its quest for adaptability remained largely unrealized.

What, they asked, should we be doing differently? We'll share with you the causes and the solutions after first reviewing the prerequisites for making breakthroughs in adaptability. As you will see, they corrected their situation and achieved adaptability by putting in place the necessary prerequisites.

THE ROUTE TO ADAPTABILITY—THE ADAPTIVE CYCLE AND ITS PREREQUISITES

Breakthroughs in adaptability create structures and processes that: 1) detect changes or trends in the internal or external environment that are potentially threatening or promising to the organization; 2) interpret and evaluate the information; and 3) refer the distilled information to empowered functions or persons within the organization who 4) take action to ward off the threats and exploit the opportunities. This is a continuous, perpetual cycle.

The cycle might more precisely be conceptualized as a *spiral*, since it goes round and round, never stopping (see Figure 10.1). Several prerequisite actions are needed to set the cycle in motion and create breakthroughs in adaptability. Although each prerequisite is essential, and all are sufficient, perhaps

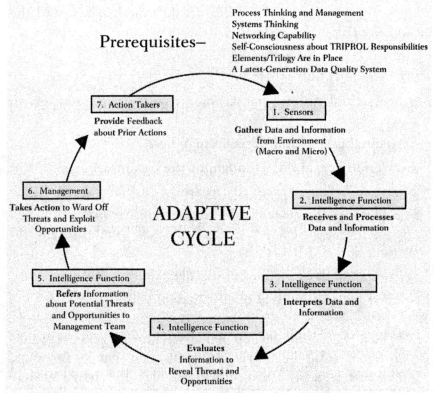

Prerequisites–

Process Thinking and Management
Systems Thinking
Networking Capability
Self-Consciousness about TRIPROL Responsibilities
Elements/Trilogy Are in Place
A Latest-Generation Data Quality System

ADAPTIVE CYCLE

7. Action Takers
Provide Feedback
about Prior Actions

1. Sensors
Gather Data and Information
from Environment
(Macro and Micro)

6. Management
Takes Action to Ward Off
Threats and Exploit
Opportunities

2. Intelligence Function
Receives and Processes
Data and Information

5. Intelligence Function
Refers Information
about Potential Threats
and Opportunities to
Management Team

3. Intelligence Function
Interprets Data and
Information

4. Intelligence Function
Evaluates
Information to
Reveal Threats and
Opportunities

FIGURE 10.1   Adaptive cycle to detect and react to organizational threats and opportunities.

the most crucial is the Data Quality Council and data quality system. All else flows from timely trustworthy data—data that purport to truthfully describe aspects of reality vital to your organization.

We will examine each prerequisite in turn.

MAKING BREAKTHROUGHS IN ADAPTABILITY HAPPEN: PREREQUISITES FOR THE ADAPTIVE CYCLE

A few organizational elements provide the structure and energy to make the adaptive cycle happen. You will recognize many of them from previous chapters. Breakthroughs in adaptability require integrating all the other breakthrough types, and the rest of the trilogy as well.

PREREQUISITES FOR THE ADAPTIVE CYCLE

- Process thinking and management, with process owners in place
- Systems thinking and decision making
- Networking capability throughout the organization
- Everyone is self-consciously aware of his/her specific triple role (TRIPROL) responsibilities as supplier, processor, and customer (see Chpater 5, "The Nature of Breakthrough" for a discussion of TRIPROL)
- All elements of the trilogy are in place and operating
- A latest-generation data quality system is in place

1. Why process thinking and management, with process owners in place? Although a typical organization may be organized and managed in vertical functional silos, the actual work of the organization, the work that ultimately results in the product or service the organization sells, is performed *cross-functionally*. This anomaly between structure and function is a basic cause of quality and performance problems as well as costs of waste, because in the absence of process owners, there typically are few if any cross-functional structures or active management. The organization in effect depends on functional management to manage cross-functional processes, with predictable, uneven results. Cross-functional waste in a functionally managed organization accounts for a significant portion of the costs of poor quality.

   When an organization is organized according to cross-functional mega-processes, it is relatively easy for the process owner, by virtue of office, to adjust procedures in the many functions through which his/her process flows in order to pull off an overall adaptive change. This is consistent with systems thinking, and it is much easier than trying to negotiate changes with each individual functional

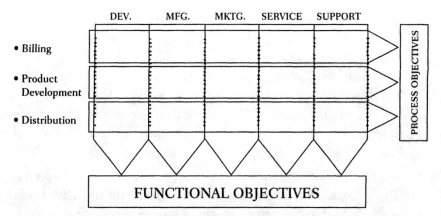

FIGURE 10.2   Functional versus process structure.

manager who may resist because he/she is in competition rather than collaboration with other functional managers. Process owners and organization members who self-consciously understand their roles in processes, in addition to understanding merely the functional tasks they must perform, can overcome these difficulties.

2. Why systems thinking? Rapid adaptive change, possibly involving the entire organization, requires a well-conceived plan that takes into account the effect a suggested change will have on each function, as well as the entire organization. That is exactly what systems thinking entails. (Please see discussion of systems thinking in Chapter 5, "The Nature of Breakthrough.") Systems thinking helps get the change right; process management helps get it fast.

3. Why networking capability? Adaptability doesn't necessarily happen from the top down. The all-powerful, all-knowing charismatic leader doesn't necessarily dream all the right dreams, and have all the brilliant flashes of intuitive insight. He/she doesn't necessarily know exactly what's going to happen and what to do about it. The world is too complex, and there's too much data and information for one person to absorb.

Consequently, the organization needs many antennae scanning the internal and external environment, and many brains communicating with each other to separate the signals from the noise. Networking provides means for communication in all directions, inside and outside of the organization. Communication is a basic element of any intelligence service. Widespread access to a computer facilitates the users' ability to search for clues, raise questions, probe deeply, verify information, solicit advice, follow suggestive trails for answers, etc.

4. Why self-consciousness about one's TRIPROL responsibilities? Both Pall and Haeckel, in their thoughtful books about adaptability refer to the commitments that various organizational role players have to each other as a thread (possibly the thread) that binds together an organization's ability to function, and particularly its ability to adapt. We refer to these commitments as TRIPROL responsibilities. TRIPROL is an acronym for "triple role." (Please see Chapter 5, "The Nature of Breakthrough," for a more thorough discussion of the TRIPROL concept.)

   Figure 10.3 shows that each organization, each organizational function, and each individual (the model applies to all these different levels) plays three contemporaneous roles: that of supplier, processor, and customer. The responsibilities in these roles are to meet certain needs of

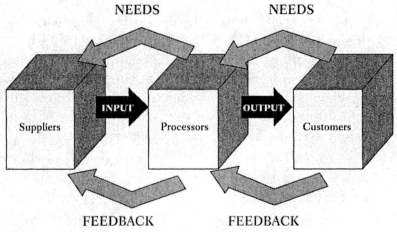

FIGURE 10.3 The TRIPROL.

their suppliers and customers. The responsibilities also include clearly communicating to others certain vital information. When the needs are not met, controls take over and adjust things so that the needs in question are met. This stable set of enforced expectations, what we call the TRIPROL and what Pall and Haeckel call commitments, are crucial not only for holding an organization together and operating repetitive processes in a consistent fashion, but they are also crucial for implementing change. To create and sustain change, one must change the expectations, the roles, and the commitments. To be successful in any role, the role player must consciously be aware of the responsibilities embedded in his/her role. Persons seeking to make adaptive change must also be aware of these same role descriptions, because the role descriptions must be changed in order to create the adaptive change.

5. Why should the elements of the Trilogy be in place? Think of it: Your organization has executed the activities and created the structures that put the Trilogy in place, and trained a sufficient number of people in the tools and techniques utilized by the trilogy. Now the organization is implementing the three major elements—the planning processes, the control processes, and the breakthrough processes. If you are a manager in this organization, you have at your command skilled personnel who have the means to respond effectively and efficiently to almost any reasonable request. This is the very nature and purpose of the trilogy, and it is a clear ingredient of perpetual adaptive change.

6. Why a latest-generation data quality system? In many aspects of organizational life, data rules. This is certainly true for adaptive breakthrough. Operational and strategic decisions are routinely based on the messages conveyed by data. Unfortunately, there is something wrong with this picture. Studies show that large percentages of data routinely used for decision making are contaminated with numerous types and generous quantities of error. Consequently, numerous wrong decisions are being made, repeatedly.

Because data are so basic and so vital to the operation of any organization, it is well worth the effort to assure that available data are accurate, precise, timely, sufficiently granular (broken down into suitably small dimensions), etc. You also need assurance that your data and information chains are under statistical control and capable of meeting required quality levels (such as <1 percent sampling error, <0.5 percent nonsampling error, etc.) If your data are this good, they can be converted to information that answers the questions being sought by gathering the data in the first place. If the data are not this good, or if you don't know how good they are, you are in the dark trying to manage something you need to see. In short, you need data you can trust; data that tells the truth and gives you reliable answers to your questions.

High-quality data can be assured with the aid of a *data quality system,* very much like the quality system your organization has in place in the production function or in other repetitive processes. Of course, a data quality system will incorporate the myriad unique quality characteristics by which data are measured. The system will also be oriented toward the various progressive activities that make up the data chain: gathering or creating the data, processing it, publishing it, using it, etc.

A data quality system originates in the Data Quality Council, which consists of appropriate members of top management, database management, IT management, quality management, database managers, etc. If elements of the data/information chain are subcontracted (say, to gather or process data), high-ranking representatives of these suppliers would also be members of the Data Quality Council.

Following is an example of the purposes and benefits of a data quality council, as published by a very large multinational manufacturer of mass consumer goods.

SOME PURPOSES AND BENEFITS OF A DATA QUALITY COUNCIL

- To provide cross-functional structure to manage cross-functional data quality activities (e.g., projects, a data quality system for all elements of the data chain)

- To assure that the data quality activities support organizational, strategic, and business goals
- To create a mechanism for upper management to provide leadership, support, and recognition in order to make cross-functional data improvement projects and systems succeed, and to nourish required cultural change
- To achieve a latest-generation data quality system that meets our internal customers' needs, and is supported by all functions and all levels

   Typical responsibilities of a data quality council are listed below.

Typical Responsibilities of a Data Quality Council

- Define and publish its responsibilities
- Formulate the data quality policy (e.g., priority of quality, need for continuous improvement, mandatory participation)
- Establish and maintain a latest-generation data quality system
- Bring all elements of the data chain into a state of statistical control
- Determine process capability of data chain processes
- Estimate the major dimensions (e.g., quality compared to competitors, quality compared to our standards, costs of poor quality, length of new data product launch cycles, etc.)
- Establish a data improvement project nomination and selection process
- Establish a project team selection process
- Provide resources: training, time for working on projects, diagnostic support, and facilitator/black belt support
- Assure that project solutions are implemented
- Establish metrics and a measuring system for tracking database quality and performance, and progress of data improvement projects

- Provide for support of projects, review of progress of projects, and coordination of projects
- Provide recognition and publicity of results
- Revise the reward system

By far the most important role of the data quality council is to establish and maintain a latest-generation data quality system. A data quality system is the totality of an organization's efforts to manage, control, and improve data quality. The system includes activities aimed at: a) understanding customer needs; b) detection and correction of specific errors; c) preventing future errors; and d) management activities to build organizational infrastructure to do so effectively and efficiently. The basic elements of a data quality system are outlined below.

BASIC (MINIMUM) OUTLINE OF THE ELEMENTS OF A DATA QUALITY SYSTEM

1. Basic definitions (quality, data quality, data quality system, process, process owner, data/information chain, data process capability, etc.)
2. Management infrastructure
   - Data quality policy (guide to management decision making)
   - Data quality council
   - Supplier management
   - Process management
   - Database of record
   - Strategic data quality management
   - Training and education
3. Technical capabilities
   - Identification of data/information chains
   - Customer needs analysis
   - Process descriptions

- Measurement
- Quality control
- Quality planning
- Quality improvement
- Inspection and test
- Quality assurance
- Document assurance
- Rewards and recognition
- Quality handbook

4. Appendix; glossary

A detailed treatment of data quality systems is the subject of an entire book or two, and consequently, is beyond the scope of this book. We would refer the reader to the published literature.

To illustrate the value of the prerequisites for adaptability described above, let's return to the true story of our troubled organization, Braggabit. Recall that they were frustrated in their quest for adaptability, and called for help. What follows is a brief description of what happened next, considerably condensed and simplified.

To start with, after an investigation into the state of current affairs, some startling discoveries were made. Among other things, the following situations came to light:

- The database managers focused almost exclusively on assembling the best data in the world. No thought was given to the ultimate customers for the data—the upper management—or their needs that the data were to satisfy. "What happens to the data I produce is not my job. My job is to get the best data." "Customers, that's Sales." "I'm doing a great job at getting the best data there are, anywhere."
- There was no agreed upon definition of "best." There certainly were huge quantities of data, broken down into seemingly infinite numbers of subcategories. But the database

managers seemed to be working essentially to meet their own needs. The notion that they had customers, much less what those customers' needs were, was a foreign concept. The database managers thought they were doing a great job.

- The data suppliers were similarly clueless about their customers' needs. They created data they thought their customers needed. After all, they were experts at creating data. Not only did the suppliers not work from clear descriptions of the Braggabit's needs, Braggabit did not even provide the suppliers with clear descriptions of Braggabit's data requirements because no clear, precise, communicable descriptions existed! The suppliers thought they were doing a great job.

- There was no unified, organized, coordinated management of the multiple databases and the multiple departmental fiefdoms that produced data. There were inadequate published quality policies to guide managerial decision making. There was, at best, an informal data quality system. For the quality of data, Braggabit depended on the individual professional competence of each database manager (who were extraordinarily qualified and competent). As a consequence of the uncoordinated, almost freelance, and somewhat competitive operation of the various departments producing data, politics associated with data issues between departments and at the overall organization level were brutal. Not really knowing where to lay blame for problems, it was reflexive to blame others. "We are doing a great job. It's the others who are causing the problems." The upper managers, who were tremendously inconvenienced and impaired by the current situation, had a tendency to blame all of them, and were greatly feared by all.

- There were several other revealing discoveries, but the ones above will suffice to make our points, below.

With this understanding of the current situation, a number of actions were taken to untangle the mess and set the operation on a unified course. Among the actions taken were:

1. *A combined supplier/customer data quality council was formed, launched, and trained.* This provided a unified

approach to managing data and information. Data quality standards became identical not only for the various data-related departments in Braggabit, but also for the outside suppliers. It also provided a forum for face-to-face discussion and joint resolution of issues, particularly supplier–customer issues, that needed immediate attention and corrective action. Most importantly, the creation of the Council established a legitimate source of policies and practices that helped get everyone to perform their jobs while operating "from the same page." Data-related politics became much more civil and much less polarized.

2. *The council members, department managers, and database managers were trained in data quality theory, particularly focusing on the TRIPROL responsibilities, and specifically each individual's TRIPROL responsibilities.* This was an "eye-opener" for many data analysts and database managers. Some for the first time came to realize and understand that the tasks they performed were in reality their contributions to larger mega-processes (data chains). They began to think of their work in terms of the roles they played in cross-functional data chains. They came to understand who their customers were and what specifically these customers needed from them. They listened to their customers, not just passively, but by actively seeking out feedback from customers about how well their needs were being met. They began to communicate specific needs to their suppliers and provide civil feedback to them about what they were doing well and what specifically needed improvement. In short, many testified that they saw themselves and their jobs in a very different light, and were energized by what they saw. Work life was good, more satisfying. Morale improved. So did the data.

3. *All parties were trained in process and systems thinking, and learned how it applied to them.* This training and the associated exercises that related the concepts discussed in the training to each trainee's job helped all participants to better understand the "big picture"—the data chain—and how all links in the chain were inter-related, and how their par-

ticular tasks affected all the other elements and were affected by them.

4. *All parties were trained in data process analysis and improvement.* Consequently, all staff became skilled in undertaking data process improvement projects. They learned how to analyze processes, diagnose causes of process problems, and install new controls to assure that: a) the new standards were met; b) improvements they devised did not disappear; and c) the gains they made were not lost. They became self-sufficient in their ability to solve data quality problems and sustain the improvements.

5. *Data improvement projects, some of them joint supplier–customer projects, were assigned by the data quality council on the basis of the biggest gaps between actual and desired performance that were identified by process analysis.* One by one, project by project, data improvement teams, many of them joint teams with members from both Braggabit and its suppliers, tackled the most pressing problems found in each step of each data chain.

6. *All elements of a latest-generation data quality system were designed, installed, and implemented, jointly, by Braggabit and its suppliers.* All data created or handled by Braggabit or its suppliers were now governed and managed to identical standards, with identical policies and practices.

PRACTICAL RESULTS OF THESE ACTIONS

The actions, as you can see, put in place most of the prerequisites for adaptability. Now Braggabit could, and did, learn in a timely fashion about what was transpiring in its business environment, and take appropriate actions to minimize its own weaknesses and exploit its competition's weaknesses and its own strengths—the very essence of adaptability (and of strategic planning, next chapter).

Take note of these observed beneficial changes:

• Upper managers were able to trust the data and information they received on a weekly basis.

- Upper management could—and did—make quicker and better marketing decisions (this was of utmost importance because Braggabit was a market-driven enterprise).

- Interdepartmental and supplier–customer relations changed from adversarial to collaborative. Gradually, mutual understanding replaced mutual suspicion. Instead of treating each other as rivals, or enemies, they treated each other as valued customers who were mutually dependent.

- Relations between database managers and upper managers also improved as they became less dubious of each other's capabilities and performance. Mutual confidence and respect was largely restored, with only a small residue of concern that lapses might occur and that people in the data chain might return to their old ways. This was a healthy concern because it motivated all those in the data chain who—from beginning to end—performed control activities to do so rigorously, without fail.

- Fewer people were diverted from their "regular" tasks and relegated to performing time-consuming and nonvalue-added incoming inspection and correction. Accordingly, head counts and costs declined, and productivity climbed.

- Having learned how to execute the elements of the Trilogy, continual improvement and adaptability became not only a possibility, but also a way of life for the organizations involved in this case.

MAKING THE ADAPTIVE CYCLE HAPPEN: THE INTELLIGENCE FUNCTION

The intelligence function is a major indispensable subsystem of any open system. It is therefore a must for any organization. Nations and armies have utilized intelligence networks for centuries to learn the capabilities, the condition, the morale, the location, the plans, etc., of the enemy. After comparing relative strengths and weaknesses of "us and them," the intelligence is utilized to determine strategies that exploit our strengths and their weaknesses.

Organizations need an intelligence function to execute every element of the trilogy. This is readily apparent when attempting breakthroughs in adaptability, which in many (but not all) ways resembles strategic planning, the subject of the next chapter. Both involve discovering events and conditions in the environment of relevance to one's organization, and executing plans that will enable the organization to prevail over its competition (in war, *enemies*) and other threats, and to pursue new promising initiatives.

One sees elements of intelligence functions in place in organizations. Universities have institutional research departments, commercial businesses have market research and field intelligence departments, and all kinds of organizations perform benchmarking to discover world-class best practices. All kinds of organizations have information systems or information technology departments.

We suggest that these various activities be combined, or at least coordinated or integrated into one repository of responsibility and authority for gathering, processing, interpreting, and evaluating data and information, and recommending to management issues that appear to indicate the need for responsive action. One might call this the intelligence department. Whatever it may be named, it will at least be a separate function, headed by someone at a director or vice-presidential level. One of its major benefits is its ability to know and retrieve all the data and information produced and stored by the organization. Management knows where to turn when it needs answers to vital questions. Furthermore, each database can be designed to be compatible in important ways with the other databases. Users can know the assumptions built into each database, how to interpret the contents, the valid and invalid uses of each, etc. Even these few benefits will make a huge improvement in organizations.

A JOURNEY AROUND THE ADAPTIVE CYCLE

Intelligence function gathers data and information from the internal and external environment. What do we need to know? Several of the following basic things, at a minimum.

From the internal environment:

- Process capability of our measurement and data systems
- Process capability of our key repetitive processes
- Performance of our key repetitive processes (human resources, sales, design, engineering, procurement, logistics, production, storage, transportation, finance, training, etc.; yields, defect types and levels, and time cycles)
- Causes of our most important performance problems
- Management instrument panel information: score cards (performance toward goals)
- Internal costs and costs of poor quality (COP^3)
- Characteristics of our organizational culture (how much does it support or subvert our goals)
- Employee needs
- Employee loyalty

From the external environment:

- Customer needs, now and in the future (what our customers or clients and potential customers or clients want from us or our products)
- Ideal designs of our products (goods, services, and information)
- Customer satisfaction levels
- Customer loyalty levels
- Scientific, technological, social, and governmental trends that can affect us
- Market research and benchmarking findings (us compared to our competition; us compared to best practices)
- Field intelligence findings (how well our products or services perform in use)

You may add to this list other information of vital interest to your particular organization. This list may seem long. It may seem expensive to get all this information. (It can be.) You may be tempted to wave it away as excessive or unnecessary. Nevertheless, if your organization is to survive, there appears to be no alternative but to gather this kind of information, and on a regular, periodic basis. Fortunately, as part of routine control and tracking procedures already in place, your organization probably gathers much of this data and information.

Gathering the rest of the information is relatively easy to justify, given the consequences of being unaware of, or deaf or blind to, vital information.

Information about internal affairs is gathered from routine production and quality reports, sales figures, accounts receivable and payable reports, monthly financial reports, shipment figures, inventories, and other standard control and tracking practices. In addition, specially designed surveys—written and interviews—can be used to gain insights into such matters as the state of employee attitudes and needs. A number of these survey instruments are available off the shelf in the marketplace. Formal studies to determine the capability of your measurement systems and your repetitive processes are routinely conducted if you are utilizing Six Sigma in your organization. Even if you don't use Six Sigma, such studies are an integral part of any contemporary quality system. Score cards are very widely utilized in organizations that carry out annual strategic planning and deployment. The scores provide management with a dashboard, or instrument panel, which indicates warnings of trouble in specific organizational areas. Final reports of operational projects from quality improvement teams, Six Sigma project teams, and other projects undertaken as part of executing the annual strategic business plan are excellent sources of "lessons learned" and ideas for future projects. The tools and techniques for conducting COP^3 studies on a continuing basis are widely available. The results of COP^3 studies become powerful drivers of new breakthrough projects because they identify specific areas in need of improvement. In sum, the materials and tools for gathering information about your organization's internal functioning are widely available and easy to use.

Gathering information about conditions in the external environment is somewhat more complex. Some approaches require considerable know-how and great care. Determining customer needs is an example of an activity that sounds simple, but actually requires some know-how to accomplish properly. Firstly, it is proactive. Potential and actual customers are personally approached and asked to describe their needs, in terms of benefits they want from a product, services, or information. Many interviewees will describe their needs in terms of a problem to be solved or a product feature. Responses like these must be translated to describe the benefits the interviewee wants, not the problem to be solved or the product feature they'd like. A detailed discussion of techniques for discovering and analyzing customer needs is found in Chapter 3, "The Planning Processes."

Tools and techniques for determining ideal designs of current and future products or services are also available. They require considerable training to acquire the skills, but the payoffs are enormous. The list of such approaches includes Quality Planning, Design for Six Sigma (DFSS), I-TRIZ, and Directed Evolution, a technique developed in Russia for projecting future customer needs and product features.

Surveys are typically used to get a feel for customer satisfaction. A "feel" may be as close as you can get to knowledge of customer feelings and perceptions. These glimpses can be useful if they reveal distinct patterns of perceptions whereby large proportions of a sample population respond very favorably or very unfavorably to a given issue. Even so, survey results can hardly be considered "data," although they have their uses, if suitable cautions are kept in mind. The limitations of survey research methodology cloud the clarity of results from surveys. (What really is the precise difference between a rating of "2" or a rating of "3"? A respondent could answer the same question different ways at 8:00 A.M. and at 3:00 P.M, for example.) (A satisfaction score increase from one month to another could be meaningless if the group of individuals polled in the second month is not the exact same group that was polled the first month. Even if they were the same individuals, the first objection raised above would still apply to confound the results.)

A more useful approach for gauging customer "satisfaction," or more precisely, their detailed responses to the products or services they get from you, is the customer loyalty study, which is conducted in person with trained interviewers every six months or so on the same people. The results of this study go way beyond the results from a survey. Results are quantified and visualized. Customers and former customers are asked carefully crafted standard questions about your organization's products and performance. Interviewers probe the responses with follow-up questions, clarifying questions, etc. From the responses, a number of revealing pieces of information are obtained and published graphically. Not only do you learn the features of your products or services that cause the respondents happiness and unhappiness, but also such things as how much improvement of defect X (late deliveries, for example) it would take for former customers to resume doing business with you. Another example: You can graphically depict the amount of sales (volume and revenue) that would result from given amounts of specific types of improvements. You can also learn what specific "bad" things you'd better improve, and the financial consequences of doing so or not doing so. Results from customer loyalty studies are powerful drivers of strategic and tactical planning, and breakthrough improvement activity.

Discovering scientific, technological, social, and governmental trends that could affect your organization simply requires plowing through numerous trade publications, journals, news media, websites, and the like, and networking as much as possible. Regular searches can be subcontracted so you receive, say, published weekly summaries of information concerning very specific types of issues of vital concern to you. Although there are numerous choices of sources of information concerning trends, there appears to be little choice of whether to acquire such information. The trick is to sort out the useful from the useless information.

A basic product of any intelligence function is to discover how the sales and performance of our organizations' products, services, and sales compare with our competitors and potential competitors. Market research and field intelligence techniques are standard features in most commercial businesses, and books on those topics proliferate.

Many organizations undertake benchmarking studies to gather information on world-class best practices. They study the inner workings of repetitive processes such as design, warehousing, operating oil wells, mail order sales, almost anything. The processes studied are not necessarily those of your competitors; they need only be the very best (efficient, effective, most economical, etc.). Benchmarking studies are classic intelligence detective work, and are often conducted on a subcontract basis with organizations who specialize in benchmarking. The results are typically published and shared with all participants. When you have discovered best practices, you can compare your performance with them and describe gaps between theirs and yours, thus identifying breakthrough opportunities.

The intelligence function

- Receives and processes data and information
- Interprets the data and information
- Evaluates the information
- Refers information about potential threats and opportunities to management

After the data are created, the structure and details of activities in the adaptive cycle, as found in your organization, will be guided by its organizational characteristics and the nature of its business. Because this is such a vital function, there will be guidelines and standards in the data quality system that embrace all important transactions within and between each step in each data chain and around the cycle. The purpose of, and justification for, the adaptive cycle is to keep management informed of whatever it needs to know about the ever-changing business environment through which they must lead the organization.

SUMMARY AND BRIDGE TO STRATEGIC PLANNING

Once the prerequisites for the adaptive cycle are in place, and once a coherent, well-managed intelligence function is operating, the organization is capable of sustained adaptability. Adaptability, as you have seen, is an integration of all the elements of the trilogy.

The contents of this chapter naturally flow into the contents of the next chapter: "Strategic Quality Planning and Deployment," which focuses on how to create ideas and plans for action and convert the ideas into actual action. Indeed, strategic planning and deployment can be conceptualized as Step 6 of the adaptive cycle: "Management takes action to ward off threats and exploit opportunities."

HIGH POINTS OF "BREAKTHROUGHS IN ADAPTABILITY"

· The survival of your organization, like all open systems, depends on its ability to quickly detect and react to threats and opportunities that present themselves from within and from outside.

· The process of quickly detecting and reacting to threats and opportunities is called "adaptability."

· Adaptability operates by means of an "adaptive cycle." The steps in the adaptive cycle are:

Sensors gather data and information from the environment

Intelligence function receives and processes data and information

Intelligence function interprets data and information

Intelligence function evaluates information to reveal threats and opportunities

Intelligence function refers information about potential threats and opportunities to management team

Management takes action to ward off threats and exploit opportunities

Action-takers provide feedback to sensors about results of action taken

- Before adaptability can be realized, and the "adaptive cycle" put into operation, a number of prerequisites must be in place. The prerequisites are:

    Process thinking and management, with process owners in place

    Systems thinking and decision making

    Networking capability throughout the organization

    Everyone is self-consciously aware of his/her specific TRIPROL responsibilities as supplier, processor and customer

    All elements of the trilogy are in place and operating

    A latest-generation data quality system

- Adaptive cycle is continuous, perpetual.

- Intelligence gathered by the adaptive cycle feeds into strategic planning and deployment (see next chapter).

C H A P T E R

E L E V E N

STRATEGIC QUALITY PLANNING AND DEPLOYMENT

Strategic quality planning (SQP) is a systematic approach for incorporating long-term business goals related to the quality of the services and goods an organization produces into the strategic plan. It also includes the deployment process as a means to achieve them.

Strategic planning has been widely used since the 1980s as a means for organizations to set a direction. Strategic planning results in an annual business plan, which includes the annual organizational budget. The strategic plan is used to plan the resources required to carry out the plan. Activities that are not in support of a plan compete for the resources that are included in the plan, and because of this, they usually are unsuccessful.

Organizations that want to institute organization-wide change programs, such as Six Sigma, that are not a part of the strategic planning process usually fail to achieve the desired results or changes. Often cited is that there were no resources available to drive the changes, and upper management did not see the initiatives as strategic. To bring attention to the importance of this activity, we refer to this process as "strategic quality planning," to make sure the executives incorporate quality-related initiatives into the annual business plan.

Organizations that want to sustain changes and want to win the support of the organization must get their activities to achieve change incorporated into the strategic plan. This will ensure the change effort will be a part of, and not have to compete with, the well-established annual business planning process. If forced to compete, it is likely that the well-intended change program or improvement efforts will result in short-term results. Organizations have managed to overcome this competitive conflict and achieve remarkable results.

The most pervasive change process of the 1980s and 1990s was total quality management (TQM), and most recently, Six Sigma. The lack of integration of TQM into the strategic plans of many organizations doomed it to failure. Most executives recall TQM as a short-term, nonbusiness–focused improvement initiative. It was not until late in the 1990s that organizations learned how to really incorporate these types of initiatives into the strategic plan. GE, Allied, and others integrated their Six Sigma initiatives into the strategic plan. This alignment with the plan, and the message articulated by its then CEO Jack Welch in his annual letter to shareholders, enabled GE to demonstrate that the Six Sigma initiative was a business and strategic initiative for GE. It therefore had a greater chance to sustain itself and get the required attention it needed to survive year after year, or until another strategy replaces it. It is also important to keep in mind that a Six Sigma change initiative is usually not the only strategy an organization must take on, but it is an important one.

This chapter describes how to integrate quality-related activities, such as a Six Sigma initiative, into the strategic planning process. It explains how SQP is managed within organizations and how they address such important issues as: how to integrate Six Sigma and the processes of the Juran Trilogy with the strategic plan and how to align strategic goals with the organization vision and mission. It also addresses the importance of how to deploy the goals throughout the organization and align improvement projects that are vital to an organization's survival.

This section:

- Discusses the means to sustain performance breakthroughs by incorporating them into the strategic planning and deployment process
- Describes the benefits of strategic planning
- Describes the systematic approach to strategic planning
- Explains the specific nondelegable roles that senior management must carry out

WHAT IS STRATEGIC PLANNING AND DEPLOYMENT?

Strategic planning and deployment is a systematic approach for setting strategies, creating goals to carry out the strategies, and converting the goals into daily, monthly, and yearly actions to meet the goals throughout the organization (Figure 11.1).

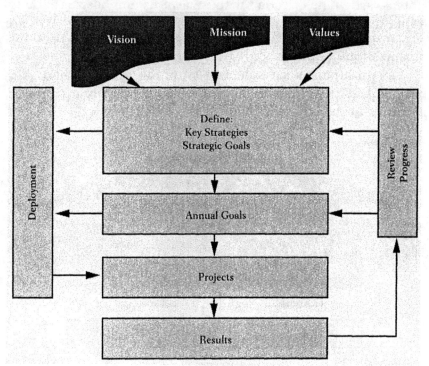

FIGURE 11.1 General planning model.

More specifically, strategic planning is a systematic process by which an organization defines its long-term goals with respect to quality, finance, human resources, sales, marketing, and R&D, and integrates them into one cohesive business plan. The plan is then deployed throughout the entire organization.

For performance breakthrough to be sustained, it must become a component of a formal organization plan to execute breakthroughs at the strategic level. The purpose of such planning is to achieve competitive advantage.

Strategic planning began to evolve during the 1980s and 1990s. It became an integral part of many organizational change processes. There were chants from executives who wanted to be number one in their industry, the best in world-class quality, the market leaders. Strategic planning also became a key element of the U.S. Malcolm Baldrige National Quality Award (Figure 11.2) and the European Foundation for Quality Management Award, as well as other international and state awards. The criteria for these awards emphasize that customer-driven quality and performance excellence initiatives are key strategic business issues that need to be integrated into the organization's strategic and annual business plans.

Organizations have achieved stunning results when they have linked their performance excellence and breakthrough activities to the strategic plans of the organization.

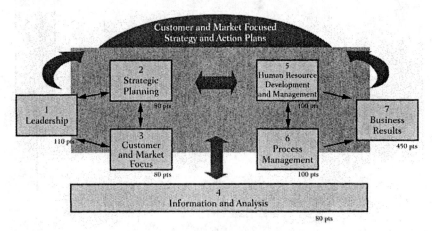

FIGURE 11.2   MBNQA award criteria framework.

Established over 30 years ago to produce and market a wide variety of consumer electronic products, Samsung Electronics has grown to become a worldwide enterprise with 58,000 employees, operating sales and marketing subsidiaries, and production factories in 46 nations. Sales for 2000 in U.S. dollars were $27.2 billion, gross profit $9.76 billion, operating profit $5.9 billion, and net income $4.78 billion—the highest since Samsung Electronics was founded.

The not so surprising major commitment Samsung Electronics is making in perfecting its fundamental approach to product, process, and personnel development is with an advanced form of the quality initiative, Six Sigma. The initiative is being used by Samsung as a process innovation tool in what appears to be record time, by producing 4,500 Six Sigma specialists, including black belts, and expects financial benefits to approach $500 million in 2001.

In 2001, plans call for a financial benefit of about $464 million through 1,577 projects, and training 582 more black belts and 3,286 green belts. In April 2001, training for 22 master black belts (MBBs) was completed. The master black belts are selected, experienced, and proven black belts who have successfully completed the rigorous MBB certification program. Going forward, Samsung's master black belts will train and coach future champion, black belt, and green belt levels.

Godfrey (1995) observed:

> To be effective, strategic planning and deployment should be used as a tool, a means to an end, not as the goal itself. It should be an endeavor that involves key personnel throughout the organization. It must capture existing activities, not just add to already overflowing plates. Finally, it must help senior managers make strategic choices, decide deliberately, set priorities, and eliminate many current activities, not just start new ones.

Prior to the 1990s, the strategic planning process in many organizations only included sales, marketing, product development, and financial goals. Goals related to Six Sigma change initiatives, or specific quality-related activities such as continuous improvement, were not included in the strategy. A structured approach was used, but it was better described as an organization-wide, financial management planning process, and it formed the basis for all business planning. This strategic planning approach was sound and often achieved its financially oriented objectives. However, it may not have been good for the customers. This method of planning consisted of only establishing financial goals, developing plans to meet the goals, providing the needed resources, establishing measures of actual performance, reviewing performance against goals, and providing rewards based on results. This planning process resulted in the annual business plan for most organizations. This plan then became the driver for all activities within the organization. Where such a plan covered a period greater than three years, it was usually referred to as a strategic business plan. A five- to ten-year horizon is strategic, but these authors have not found many organizations looking beyond five years, due to the rapid changes in our society.

As we have now seen companies like GE, Dow Chemical, and others looking at Six Sigma as a long-term strategy, we have seen the incorporation of these as key strategies in their plans. By adding strategic goals related to Six Sigma and quality, this financially focused plan has become customer-driven, or in our words, a strategic quality plan.

The major components of the traditional financial planning process that were effective were:

• A *hierarchy of goals*. This includes, but is not limited to financial goals supported by a hierarchy of financial goals at lower levels: divisional and departmental budgets, sales quotas, cost standards, project cost estimates, etc.
• A *formalized methodology* for establishing the goals and for providing the needed resources to achieve the strategic plan or annual business plan.
• An *infrastructure* which (usually) includes a finance committee, a full-time controller and supporting personnel, and all top management.

- A *control process* that includes: systems for data collection and analysis, financial reports, reviews of financial performance against goals, and market share measures.
- *Provision of rewards.* Performance against financial goals is given substantial weight in the system of merit rating and recognition of key employees.
- *Universal participation.* The financial goals, reports, reviews, etc., are designed hierarchically to parallel the organization's organization hierarchy. These hierarchical goals make it possible for managers at all levels to support upper managers in managing for finance.
- *A common language.* The planning process typically focuses on major, common metrics "revenues and profits"—expressed in a common unit of measure—a currency unit, such as the U.S. dollar. There are also other common metrics which are widely used; ratios such as return on investment and return on sales are examples. In addition, such key words as budget, expense, profit, etc., acquire standardized meanings, so that communication becomes more and more precise. Hence, the organization creates a language it can understand. Six Sigma itself is a set of common terms and measures for quality.

The approach used to establish company-wide financial management and incorporate it into the strategic plan is applicable to the establishment of any organization-wide change effort. The generic steps and features inherent in managing for the annual business plan are likewise applicable to managing for performance breakthrough. It also makes it easier to incorporate organization-wide improvement programs into one cohesive plan. To incorporate Six Sigma change initiatives and quality-related goals into the plan, the organization need not learn a new process. It simply needs to add the goals related to the Six Sigma initiative or change initiative desired.

Strategic "quality" planning must include:

- A *customer-focused vision*, not just a vision focused on financial results.

- *Six Sigma or quality-related goals.* The major quality goals are incorporated and are supported by a hierarchy of goals at lower levels in the form of subgoals, projects, and so on. The strategic quality plan incorporates the voice of the customer with quality goals and integrates them throughout the plan. This integration enables the goals to be legitimate and balances the financial goals (important to shareholders) with those of importance to the customers. It also eliminates the concern that there are two plans, one for finance and one for quality!

- *A formalized methodology*, a systematic structured process for establishing goals and providing resources.

- *A new infrastructure* is created, which includes the establishment of an upper management team or "executive council" and supporting personnel (Figure 11.3).

- *A review and control process*, or more commonly referred to as a balanced score card (Kaplan), which includes systems for data collection and analysis of customer data, as well as quality reports and reviews of key performance indicators to monitor performance against goals.

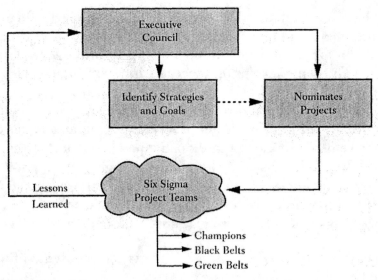

FIGURE 11.3   Infrastructure for change.

- *Provision of rewards.* Performance against quality goals is given substantial weight in the system of merit rating and recognition. A change in the structure that includes rewarding the right behaviors is required.
- *Universal participation.* The goals, reports, reviews, etc. are designed to gain participation from the organization's hierarchy. This participation makes it possible for employees at all levels to support upper managers in managing for quality.
- *Training.* It is common for all employees to undergo training in various quality concepts, processes, methods, and tools. Organizations, which have so trained their work force, in all functions and at all levels, are well positioned to outperform organizations in which such training has been confined to the quality department or managerial employees.

> Spending on Six Sigma training and projects, including certification of employees as "Green Belts, Black Belts, and Master Belts" in Six Sigma, is up from $200 million in 1996 to $450 million this year. (CEO Jack Welch) counts annual benefits as rising from $170 million to $1.2 billion. Six Sigma alone will add 25 cents to earnings per share in 1998 and $1 by 2000. GE earned $2.50 a share in 1997.
>
> —*USA Today*, February 27, 1998

The changes required are obviously numerous, but usually not extensive. Prior to the 1980s the asserted benefits of strategic planning were generally not persuasive to upper managers. Most of the reasons are implied in that same list of changes:

- "Going into strategic planning looks like a lot of work."
- "It adds to the work load of upper managers, as well as lower levels."
- "It is quite disturbing to the established cultural pattern."
- "We've already tried it and it failed."
- "We cannot think that far ahead; we are worried about tomorrow."

WHY DO STRATEGIC QUALITY PLANNING?

The first question is: Why do it? Can it help us achieve our business results and become a global leader? To answer these questions requires a look at the benefits that other organizations have realized from SQP. We reviewed a number of organizations and they report that SQP:

1. Focuses the organization's resources on the activities that are essential to increasing customer satisfaction, lowering costs, and increasing shareholder value

2. Creates a planning and implementation system that is responsive, flexible, and disciplined

3. Encourages interdepartmental cooperation (because most all processes that impact customers are multifunctional)

4. Provides a process to execute breakthroughs year after year

5. Empowers managers and employees by providing them with the authority to carry out the planned activities (because they were part of the planning process)

6. Eliminates unnecessary and wasteful team activities that are not in the plan

7. Eliminates the existence of two potentially conflicting plans: the finance plan and the Six Sigma plan

8. Focuses resources to assure financial plans are achievable

Many different companies have tried to implement change programs based on Six Sigma and TQM. Some organizations have achieved stunning results; others have been disappointed by their results, often achieving little in the way of bottom-line savings or increased customer satisfaction. Some of these efforts have been classified as failures.

One of the primary causes of these disappointments has been the failure to incorporate these quality "programs" into the business plans of the organization.

Other reasons for failure:

1. Strategic planning was assigned to planning departments, not to the upper managers themselves. These planners lacked training in quality concepts and methods and were not among the decision makers in the organization. This led to a strategic plan that did not address the impact of quality on business results.

2. Individual departments had been pursuing their own departmental goals, failing to integrate them with the overall company goals.

3. New products or services continued to be designed with failures from prior designs that were carried over into new models, year after year, because the new designs were not customer-driven or did not incorporate goals and strategies defined in the organization-wide plan.

4. Multifunctional "re-engineering" projects such as SAP or large ERP implementation have suffered delays and waste, due to inadequate participation and to lack of early warnings by upper management, and have ended before positive business results were achieved.

5. There has been no clear responsibility for reducing cycle times or waste associated with major business processes. Clear responsibilities are limited to local (intradepartmental) processes, but these processes cut across many departments.

6. Quality goals were assumed to apply only to manufactured goods and manufacturing processes. Customers became irritated not only at receiving defective goods, but also at receiving incorrect invoices and late deliveries. The business processes that produce invoices and deliveries were not subjected to modern quality planning and improvement because there were no goals in the annual plan to do so.

The deficiencies of the past had their origin in a lack of a systematic, structured approach such as the one that already exists in managing for finance. The existence of this approach then led logically to a proposal for a remedy: a remedy that treats managing for quality on the same organization-wide basis as managing for finance. The remedy requires incorporating Six Sigma initiatives into the strategic plan.

THE STRATEGIC QUALITY PLANNING TERMINOLOGY

Strategic planning and deployment process requires that an organization incorporate customer-focused quality into the organization's vision, mission, values, policies, strategies, long- and short-term goals, and projects. Projects are the day-to-day, month-to-month activities that link quality planning and quality improvement to the organization's business objectives.

The elements needed to establish SQP are generally alike for all organizations. However, each organization's uniqueness will determine the sequence and pace of application and the extent to which additional elements must be provided.

There exists an abundance of jargon used to communicate the strategic planning and deployment process. Depending on the organization, one may use different terms to describe similar concepts. For example, what one organization calls a vision, another organization may call a mission.

The following definitions are in widespread use and are used in this section:

- *Vision.* A desired future state of the organization. Imagination and inspiration are important components of a vision. Typically, a vision can be viewed as the ultimate goal of the organization, one that may take five years or even ten years to achieve.

- *Mission.* The purpose or reason for the organization's existence, (i.e., what we do and who we serve).

- *Values.* What the organization stands for and believes in.

- *Strategies.* A means to realize the vision. Strategies are few and define the key success factors such as price, value, market share, and culture that the organization must pursue. Strategies are sometimes referred to as "key objectives" or "long-term goals."

- *Goals.* What the organization must achieve over a one- to three-year period; the aim or end to which work effort is directed. Goals are referred to as "long-term" (two to three years) and "short-term (one to two years). Achievement of goals signals the successful execution of a strategy.

SIX SIGMA TESTIMONIAL

Out of the recent merger of Allied Signal and Honeywell came the company we now call Honeywell. Its CFO, Richard Wallman, who came from Allied Signal, has had a number of best practices ready to roll across the new combined company.

One of them involved the application of Six Sigma to administrative (transactional) processes. Prior to the merger, Wallman implemented Six Sigma across Allied Signal's revenue chain—from order acquisition to cash collection—to address the company's need for better working capital management.

The outcome has been impressive. When cash management solutions, such as credit card payments, electronic fund transfers, and web-based pricing improvements were implemented with Six Sigma methods, they reduced "past dues" by 38 percent and delivered $100 million in cash flow over the first two years.

Wallman stated, "This initiative pays a lot of dividends to the bottom line. We created enhanced customer relations, higher employee satisfaction, a five-day reduction in DSO, and the liberation of $700 million of cash flow over the past three years."

—*2001 CFO Magazine*

- *Projects.* Medium-term (3–12 months) activities that address a deployed annual goal, and whose successful completion contributes to assurance that the higher-level subgoal and strategic goal are achieved. A project most usually implies assignment of selected individuals to a team that is given the responsibility and authority to achieve the specific annual goal. Six Sigma Improvement and Design for Six Sigma projects are examples of projects that must be aligned to the plan.

- *Deployment.* To turn a vision into action, the vision must be broken apart and translated into successively smaller and more specific parts (i.e., from vision to key strategies, strategic goals, subgoals, annual goals, and down to the level of projects and even departmental action). The detailed plan for this decomposition and distribution throughout the organization is called the "deployment plan." It includes the assignment of roles and responsibilities, and identification of resources to implement the necessary activities and achieve the goals.

- *Key performance indicators.* The measurements that are visible throughout the organization for evaluating the degree to which the strategic plan (strategic goals, subgoals, and annual goals) is achieved. They make up the balanced score card in many organizations (Figure 11.4).

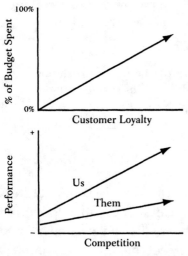
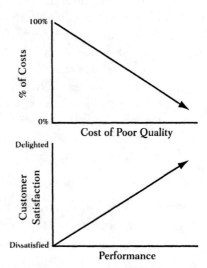

FIGURE 11.4   Key quality goals.

THE BASIC ELEMENTS OF STRATEGIC PLANNING: A VISION

SQP usually begins by assuring the organization's statement of a vision defines a benefit the customer will achieve when the organization achieves the vision. To be successful, the vision should express the benefit in a way that the customers and shareholders see the value in it. A vision should define the benefit a customer, an employee, a shareholder, or society can expect from the organization. A traditional vision statement of the past sounded very internally driven such as, "we want to have the largest market share in our industry," "we want to be a billion euro organization by 2005," and so on. These are nice, but to whom? The customer may not see the benefit of your goal. Visions that matter to customers are described as follows:

CUSTOMER-FOCUSED VISIONS

- "We will lead in delivering affordable, quality health care that exceeds the service and value of our customers." —*Kaiser-Permanente*
- "We believe our first responsibility is to our doctors, nurses, and patients, to mothers and fathers, and all others who use our products and services...." Our Credo—*Johnson & Johnson*
- "To engineer, produce, and market the world's finest automobiles."—*Cadillac Motor Car*

Each of the above visions offers a very different view on the direction and character of the organization. Each conveys a general image to customers and employees of where the organization is headed. For the organization, it is often the first time in its history it has had a clear picture of where it is headed, why it is going there, and what activities are necessary to take it there.

Good vision statements are also doable, compelling, and shared throughout the organization. It is often a good idea to make the vision a stretch goal, within a three- to five-year framework, and state a measurable achievement (i.e., being the best). In creating the vision, an organization should take into account the customers, the market(s) in which it wants to compete, the environment within which the organization operates, and the current state of the organization's culture.

Vision statements, by themselves, are little more than words. Publication of such statements does not inform the members of an organization what it should do differently from what it has done in the past. The strategic planning and deployment process and the strategic plan become the basis for making the vision a reality. The words of the vision are just a reminder of what the organization is pursuing. The vision must be carried out through deeds and action.

Some common pitfalls in forming a customer-driven vision are:

- Failure to involve key executives at all levels in creating the vision
- Failure to consider the effects that the rapid changes taking place in the global economy will have three to five years in the future on your plans
- Failure to benchmark competitors or to consider all possible sources of information on future needs, internal capabilities, and external trends—including suppliers
- Creating a vision that is too easy or too difficult to achieve
- Failure to explain the vision as a benefit to customers, employees, suppliers, and other stakeholders
- Thinking that once a strategic plan is written, it will be carried out with no further work
- Focusing the vision exclusively on shareholders as customers

THE BASIC ELEMENTS OF STRATEGIC PLANNING: A MISSION

A mission is often confused with a vision, and even published as one. A mission statement is designed to address the question, "What business(es) are we in?" A mission should clarify the organization's purpose or reason for existence. That is all. The following are some examples.

EXAMPLES OF MISSION

- "The Ritz-Carlton Hotel is a place where the genuine care and comfort of our guests is our highest mission."—*Ritz-Carlton Hotel*
- "We exist to create, make, and market useful products and services to satisfy the needs of its customers throughout the world."—*Texas Instruments*
- "Our mission is to be a leader in meeting the present and future health care needs of the people of our communities through teaching, research programs, and a network of high-quality services, which share common goals and values." —*Sentara Health System*

Together, a vision and a mission provide a common agreed-upon direction for the entire organization. This direction can be used as a basis for daily decision making.

THE BASIC ELEMENTS OF STRATEGIC PLANNING: DEVELOPING KEY STRATEGIES

The first step in converting the vision into an achievable long-term plan is to break the vision into a small number (usually five) of key strategies. Key strategies represent the fundamental choices that the organization will need to make on how it will go about reaching its vision. Each strategy must contribute significantly to the overall vision. For example:

EXAMPLES OF KEY STRATEGIES

At Cadillac, the vision "to engineer, produce, and market the world's finest automobiles" led to three critical strategies implemented to transform Cadillac. These strategies are:

- A cultural change where teamwork and employee involvement are considered a competitive advantage
- A focus on the customer with customer satisfaction in the master plan
- A more disciplined approach to planning that focuses all employees on the quality objectives

In order to determine what the key strategies should be, one needs to assess five areas of the organization and obtain the necessary data on:

- Customer and client satisfaction and loyalty
- Costs related to poor performing processes
- Employee satisfaction
- Internal business and quality systems (including suppliers)
- Competitive benchmarking

Each of these areas for assessments can form the basis for developing a balanced quality scorecard as well. Setting key strategies requires specific data on the quality position and environment. These data must be analyzed to discover specific strengths, weaknesses, opportunities, and threats as they relate to customers, quality, and costs. They therefore become strategies and metrics on the scorecard.

THE BASIC ELEMENTS OF STRATEGIC PLANNING: DEPLOYING STRATEGIC QUALITY GOALS

Depending on the size of the organization, it may need to set specific, measurable strategic goals that must be achieved for the broad strategy to be a success. These quantitative goals will guide the organization's efforts toward achieving each strategy. As used here, a goal is an aimed-at target. A goal must be as specific as it can be. It should be quantified (measurable) and achieved within a specific period. At first, an organization may not know how specific the goal should be. Over time, the measurement systems will improve and the goal setting will become more specific and more measurable. The following is an example of a measurable key quality goal:

"A CHANGE IN STRATEGY FOR DOW"

"We expect Six Sigma to elevate our company to an entirely new level of operational performance, delivering $1.5 billion in EBI cumulatively by 2003 from the combined impact of revenue growth, cost reduction, and asset utilization."
—1999 TDCC *Annual Report*

WHAT WILL SIX SIGMA DO FOR DOW?

"Six Sigma will be a vehicle to transform this company to Premier Status: in the eyes of our competitors, in the eyes of Wall Street and, at the very foundation of our company, in the eyes of our employees. We will use it to drive increased customer loyalty, better bottom line results, and to reduce employee frustration over rework, broken processes, and poor quality."

Bill Stravopoulos
Chairman of the Board
The Dow Chemical Company

Despite the uniqueness of specific industries and organizations, certain subjects for quality goals are widely applicable. There are five areas that are minimally required to assure the proper goals. They are:

- *Product and service performance.* Goals in this area relate to product features that determine response to customer needs: promptness of service, fuel consumption, mean time between failures, and courtesy. These features directly influence product salability and impact company revenues when they are met.
- *Competitive performance.* This has always been a goal in market-based economies, but seldom a part of the business

plan. The trend to make competitive quality performance a part of the business plan is recent, but irreversible. It differs from other goals in that it sets the target relative to the competition, a rapidly moving target. For example: Our products will be considered the best in class within one year of introduction versus a typical goal that our product will meet our financial or sales projections in its first year.

- *Sigma value and quality improvement.* Goals in this area are aimed at improving product salability by reducing the cost of poor performing processes so that the price to customers is the best in class. Either way, the goal is deployed through a formal list of Six Sigma improvement projects with associated assignment of responsibilities. Collectively, the projects focus on reducing deficiencies in the organization.

- *Cost of poorly performing processes (COP³).* Goals related to quality improvement usually include a goal of reducing the costs due to poor quality or waste in the processes. These costs are not known with precision, though they are estimated to be very high. Nevertheless, it is feasible, through estimates, to bring this goal into the business plan and to deploy it successfully to lower levels. A typical goal might be "to reduce COP³ 50 percent from this year's level within three years."

We have defined the cost of poor quality as those costs that would disappear if our products and processes were perfect and generated no waste. Those costs are huge. As of the 1980s, about a third of the work in the U.S. economy consisted of redoing prior work because products and processes were not perfect.

The costs are not known with precision. In most organizations, the accounting system provides only a portion of the information needed to quantify this cost of poor quality. It takes a great deal of time and effort to extend the accounting system to provide full coverage. Most organizations have concluded that such effort is not cost-effective. What can be done is to fill the gap by estimates, which provide upper managers with approximate information as to the total cost of poor quality, and as to which are the major areas of concentration. These concentrations then become the target for quality improvement projects. Thereafter, the completed projects do provide fairly precise figures on quality costs before and after the improvements.

- *Performance of key macro-business processes.* Goals in this area have only recently entered the strategic plan. As more organizations manage with process owners, the goals that relate to the performance of major processes, which are multi-functional in nature, (e.g., new product development, supply chain management, and associated conversion process[es], and processes such as accounts receivable and purchasing are entering the plan). For such macro-processes, a special problem is to decide who should have the responsibility for meeting what goal. Many organizations are frustrated when goals are not met. One reason is that the goal deployment of multifunctional processes cannot be assigned to a functional leader to achieve them. Multifunctional goals must be assigned to multifunctional teams (Figure 11.5). The leader of this may be a black belt or champion.

- *Customer satisfaction.* Setting specific goals for customer satisfaction helps keep the organization focused on the customer. Clearly, deployment of these goals requires a good deal of sound data on the current level of satisfaction and what

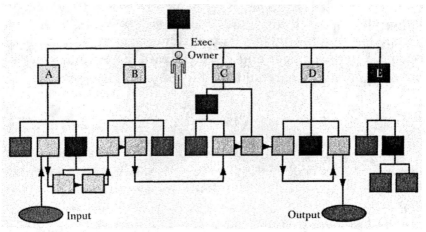

FIGURE 11.5   Macroprocesses cut across multiple departments.

factors will contribute to increasing satisfaction and removing dissatisfaction. If the customers' most important needs are known, the organization's strategies can be targeted to meet those needs effectively.

- *Customer loyalty and retention.* Beyond direct measurement of customer satisfaction, it is even more useful to understand customer loyalty. A goal stated as "we want to achieve 95 percent retention rate of our 2003 customers working with us in 2004" sets the direction that we must do whatever we can to keep our customers with us. This may translate into the total dollars each customer spends each year with your organization. The organization can benchmark to discover the competition's performance, then set loyalty goals to beat that performance.

The goals selected for the annual business plan are chosen from a list of nominations made at the highest levels of the organization. Only a few of these nominations will survive the screening process and end up as part of the organization-wide business plan. Other nominations may instead enter the business plans at lower levels in the organization. Many nominations will be deferred because they were unable to secure the necessary priority and, therefore, will get no organization resources.

Upper managers should become an important source of nominations for strategic quality goals, since they receive important inputs from sources such as membership on the executive council, contacts with customers, periodic reviews of business performance, and contacts with upper managers in other companies, shareholders, and employee complaints.

LAUNCHING A STRATEGIC QUALITY PLAN

Creating a strategic plan that is customer-focused requires that leaders become coaches and teachers who are personally involved, are consistent, eliminate the atmosphere of blame, and make their decisions on the best available data. Juran (1999) has stated: "You need participation by the people that are going to be impacted, not just in the execution of the plan, but in the planning itself. You have to be able to go slow, no surprises, use test sites in order to get an understanding of what are some things that are damaging, and correct them." If this simple lesson is followed, more organization's strategic plans will be achieved.

UPPER MANAGEMENT LEADERSHIP

A fundamental step in the establishment of any strategic plan is the participation of upper management on an executive council (Figure 11.6). Membership typically consists of the key executives. Management usually comes together as a team (if it is not already) to determine and agree upon the strategic direction of the organization. It is recommended that a "council" be formed to oversee and coordinate all strategic activities aimed at achieving the strategic plan. The council is responsible for executing the strategic business plan and monitoring the key performance indicators.

The executive council is also responsible for assuring that similar councils or steering teams are established at the business unit level and they, in turn, should establish councils at the next level (for example, at the respective divisional and facility [site] levels). In such cases, the councils are interlocked

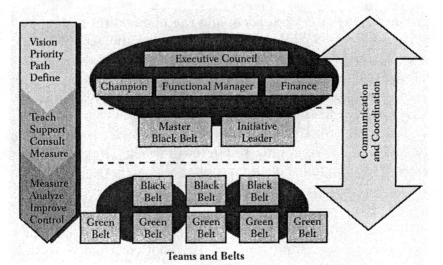

FIGURE 11.6   Six Sigma infrastructure.

(i.e., members of upper level councils serve as chairpersons for lower level councils).

If a council or management team similar to it is not in place, the organization will have to create one. In most companies, especially global organizations, many processes are too complex to be managed functionally. A council assures a multifunctional team working together to maximize process efficiency and effectiveness. Although this may sound easy, in practice, it is not. The senior management team members may not want to give up the "monopolies" they have enjoyed in the past. For instance, the manager of sales and marketing is accustomed to defining customer needs, the manager of engineering is accustomed to sole responsibility to create products, and the manager of manufacturing has enjoyed free rein in producing products. In the short run, these managers may not easily give up their monopolies to become team players.

Deploying the goals to lower levels. Deployment is the conversion of goals into operational plans and projects. "Deployment" as used here means subdividing strategic goals into subgoals, subgoals into annual goals, and assigning annual goals to lower levels of the organization in the form of projects. This conversion requires careful attention to such details as the actions needed to meet these goals, who is to take these actions, the resources needed, and the planned timetables and milestones. Successful deployment requires the establishment of an infrastructure for managing to the plan. It also requires interlocking councils throughout the organization, along with putting project teams in place. Goals are deployed to multifunctional teams, functions, and individuals.

Once the strategic goals have been agreed to, they must be subdivided and communicated to lower levels. The deployment process also includes dividing up broad goals into manageable pieces (short-term goals or projects). For example:

- An airline goal of attaining 95 percent on-time arrivals may require specific short-term (8–12 months) projects to deal with such matters as:

The policy of delaying departures in order to accommodate delayed connecting flights

The organization for decision making at departure gates

The availability of equipment

The need for revisions in departmental procedures

The state of employee behavior and awareness

- A hospital's goal of improving the health status of the communities they serve may require projects that:

Reduce incidence of preventable disease and illness

Improve patient access to care

Improve the management of chronic conditions

Develop new services and programs in response to community needs

Such deployment accomplishes some essential purposes:

- The subdivision continues until it identifies specific deeds to be done.
- The allocation continues until it assigns specific responsibility for doing the specific deeds.

Those who are assigned responsibility respond by determining the resources needed and communicating this to higher levels. Many times the council must define specific projects, complete with team charters and team members, to assure goals are met (Figure 11.7).

Communicating the plan. Once the goals have been established, the goals are communicated to the appropriate organization units. In effect, the executive leadership asks their top management, "What do you need to support this goal?" The next level managers discuss the goal and ask their subordinates a similar question, and so on. The responses are summarized and passed back up to the executives. This process may be repeated several times until there is general satisfaction with the final plan.

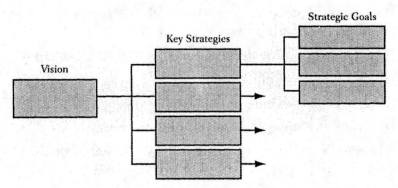

FIGURE 11.7   Strategic goals—vision converted to operational plans.

Successful deployment requires an infrastructure for managing quality. (See Chapter 8, "Breakthroughs in Current Performance.") This infrastructure includes project teams in place and, in a large organization, interlocking councils throughout the organization.

To some degree, deployment can follow hierarchical lines: corporate to division, division to function, etc. However, this simple arrangement fails when goals relate to cross-functional business processes and problems that affect customers.

Major activities of organizations are carried out by use of interconnecting networks of business processes. Each business process is a multifunctional system consisting of a series of sequential operations. Being multifunctional, the process has no single "owner," hence no obvious answer to the question: Deployment to whom? Deployment is thus made to multifunctional teams. At the conclusion of the team project, an owner is identified. the owner then monitors and maintains this business process.

The tree diagram is a graphic tool that aids in the deployment process (see Figure 11.8). A tree diagram displays the hierarchical relationship among the vision, key strategies, strategic goals, subgoals, annual goals, and projects. A tree diagram is useful in visualizing the relationship between goals and project objectives. It also provides a visual way to determine if all strategies are supported.

Reviewing progress and a note on balanced scorecards.
To ensure that all employees understand the importance of the

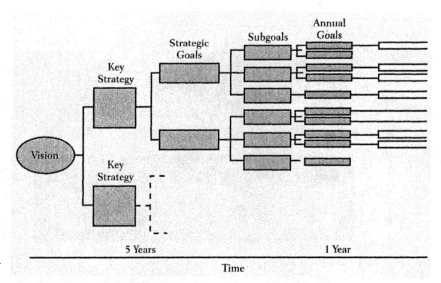

FIGURE 11.8   Strategic deployment.

strategic plan, executives must periodically review its progress. Measurement of performance is necessary and there should be an organized approach to it. Organizations that have a balanced scorecard understand the benefits of monitoring key indicators and re-allocating resources if the goals are not being achieved. They have found that:

- Performance measures indicate the degree of accomplishment of objectives and, therefore, quantify progress toward the attainment of goals.
- Performance measures are needed to implement the process of Six Sigma and continuous improvement, which is central to the changes required to become customer-focused.
- Measures of individual, team, and business unit performance are required for periodic performance reviews by management.

A formal, efficient review process will increase the probability of reaching goals. When planning actions, an organization should look at the gaps between the current state and the goal it is seeking. The review process looks at gaps between what has been achieved and what is the goal (Figure 11.9).

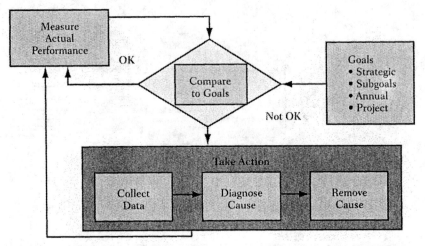

FIGURE 11.9 Strategic feedback loop.

Frequent measurements of attainment of goals (annual goals, subgoals, and strategic goals) displayed in graphic form helps identify the gaps in need of attention. Success in closing those gaps depends on a formal feedback loop, with clear responsibility and authority for acting on those differences. In addition to the review of results, progress reviews are needed for projects underway to identify potential problems before it is too late to take effective action. Every project should have specific, planned review points, much like those in Figure 11.10.

Once the goals have been set and broken down into subgoals, key measures (performance indicators) need to be established. A measurement system that clearly monitors performance to goals has the following properties:

- The indicators link strongly to strategic goals and to the vision and mission of the organization.
- The indicators reflect customer concerns; that is, the measures focus on the needs and requirements of internal and external customers.
- The measurement system provides data on a small number of "key" measures on key processes that is accessible on a timely basis for executive decision making.

Project	Project Leader	Baseline Measurements	Targets	Review Points						Review Leader
				Initial Plan	Resources	Analysis	Plan	Results		

FIGURE 11.10   Progress review plan.

• The measurement system identifies chronic wastes and the efforts needed to reduce these costs and improve efficiencies.

For example: The best measures of implementation of the strategic plan are simple, quantitative, and graphical. A basic spreadsheet which describes the key measures and how they will be implemented is shown in Figure 11.11.

As goals are set and deployed, the means to achieve them at each level must be analyzed to ensure that they satisfy the objective that they support. Then the proposed resource expenditure must be compared with the proposed result and the benefit–cost ratio assessed.

Examples of measures include:

• Financial results:

Gains

Investment

Return on investment

• Defect rate reductions, as measured in sigma values
• New product development:

Number or percentage of successful product launches

Annual Goals	Specific Measurements	Frequency	Format	Data Source	Name

FIGURE 11.11   Measurement of annual goals.

Return on investment of new product development effort

Cost of developing a product

Percent of revenue attributable to new products (less than two years old)

Percent of market share gain attributable to products launched during last two years

Percent of on-time product launches

Cost of poor quality associated with new product development

· Supply chain management:

Manufacturing lead times

Inventory turnover

Percent of on-time delivery

First pass yield

Cost of poor quality

The following is an example of measures used by Banc One for its tellers:

Speed

· Number of customers in the queue

· Amount of time in the queue

Timeliness
- Time per transaction
- Turnaround time for no-wait or mail transactions

Accuracy
- Teller differences
- Amount charged off/amount handled

To enable upper managers to "know the score" relative to achieving the strategic plan, it is necessary to design a report package. In effect, the choice of subjects identifies the instruments needed on the upper management instrument panel. The system of reports is what connects those instruments to the data sources.

The report package consists of several conventional components:

- Quantitative reports on performance, based on data
- Narrative reports on such matters as threats, opportunities, pertinent events
- Audits conducted (see below)

These conventional components are supplemented as required to deal with the fact that each organization is different. The result should be a report package that assists upper managers in meeting the specified quality goals in much the same way as the financial report package assists upper managers in meeting financial goals.

The council has the ultimate responsibility for the design of such a report package. In large organizations, the design of such a report package requires inputs from the corporate offices and divisional offices alike. At the division level the inputs should be from multi-functional sources.

The report package should be specially designed to be read at a glance, and to permit easy concentration on those exceptional matters that call for attention and action. Reports in tabular form should present the three essentials: goals, actual performances, and variances. Reports in graphic form should show minimally the trends of performances against goals. The choice of format should be made only after learning what are the preferences of the customers (i.e., upper managers).

Managerial reports on quality are usually published monthly or quarterly. The schedule is established to coincide with the meeting schedule of the quality council or other management review body.

The editor of the quality report package is usually the director of quality (quality manager, etc.), who is usually also the secretary of the quality council.

In Texas Instruments, Inc., the quality report package (the "Quality Blue Book") was deliberately designed to parallel the organization's financial reporting system, down to the color of the cover (blue). The report is organized into:

1. Leading indicators (e.g., quality of purchased components)
2. Concurrent indicators (e.g., product test results; process conditions; service to customers)
3. Lagging indicators (e.g., data feedback from customers; returns)
4. Data on cost of poor quality

The report is issued monthly, and is the basis for annual performance appraisal of managers' contributions to quality.

The reports, both quantitative and narrative, should be reviewed formally on a regular schedule. Formality adds legitimacy and status to the reports. Scheduling the reviews adds visibility. The fact that upper managers personally participate in the reviews indicates to the rest of the organization that the reviews are of great importance and that management is seriously committed to organization-wide improvement.

These audits may be based on externally developed criteria, specific internal objectives, or some combination of both. Three well-known external sets of criteria are those of the United States' Malcolm Baldrige National Quality Award, the European Quality Award, and Japan's Deming Prize.

Traditionally, quality audits have been used to provide assurance that products conform to specifications and that operations conform to procedures. At upper management levels, the subject matter of quality audits should expand to provide answers to such questions as:

- Are our policies and goals appropriate to our organization's mission?
- Does our quality provide product satisfaction to our clients?
- Is our quality competitive with the moving target of the marketplace?
- Are we making progress in reducing the cost of poor quality?
- Is the collaboration among our functional departments adequate to assure optimizing company performance?
- Are we meeting our responsibilities to society?

Questions such as these are not answered by conventional technological audits. Moreover, the auditors who conduct technological audits seldom have the managerial experience and training needed to conduct business-oriented quality audits. In consequence, organizations that wish to carry out quality audits oriented to business matters usually do so by using upper managers or outside consultants as auditors. The widest use of this concept has been in the major Japanese organizations.

Juran has stated: "One of the things the upper managers should do is maintain an audit of how the processes of managing for achieving the plan is being carried out. When you go into an audit, you have three things to do. One is to identify what are the questions to which we need answers. That's nondelegable; the upper managers have to participate in identifying these questions. Then you have to put together the information that is needed to give the answers to those questions. That can be delegated and that is most of the work, collecting and analyzing the data. In addition, there are the decisions of what to do in light of those answers, which are nondelegable. That is something the upper managers must participate in."

Audits by upper managers should be scheduled with enough lead time for preparing the needed information base. The subject matter should likewise be determined in advance, based on prior discussions by the quality council. Some of these audits are conducted on-site at major facilities or regions. In such cases, local managers are able to be active participants through making presentations, responding to questions, guiding the upper managers during the tour of the facility, and so on.

The financial, human resource, and quality officers, those executives that manage the change initiatives, will have to monitor goals and the strategy, and bring that strategy to life throughout the organization's business cycle.

Strategic planning and deployment is a systematic approach for integrating customer focus and organization-wide quality improvement with the strategic plans throughout the entire organization. The strategic planning and deployment process provides focus and enables organizations to align improvement goals and actions with the vision, mission, and key strategies. Strategic planning provides the basis for senior management to make sound strategic choices and prioritize the organization's improvement and other change activities. Activities not aligned with the organization's strategic goals should be terminated or eliminated. All too often, organizations are deluged with too many activities, projects, and change efforts. The necessary investments in time, money, and resources should be allocated, and the assigned personnel should be provided the necessary time and right motivations to ensure successful achievement of specified goals.

HIGH POINTS OF "STRATEGIC QUALITY PLANNING AND DEVELOPMENT"

- Strategic quality planning and deployment is a systematic approach for setting strategies that will assure survival by creating goals to realize the strategies and converting the goals into daily, monthly, and yearly actions to meet the goals throughout the organization.

- The results of strategic quality planning and deployment are a business plan, a budget, and a *unified* organization whose members understand the mission, vision, strategies, and what each individual is expected to do to reach goals and how progress to goals will be measured.

A ROAD MAP
FOR CHANGE

We have come to the last chapter in our book. Although it is last in our book, it should be the first chapter to begin your strategic journey to transform your organization, while reaping the benefits of that transformation.

The Juran road map for change is a systematic (but not prescriptive) approach for going from "here to there." It will provide a guide for incorporating long-term business goals related to the intended change with the organizational resources required to carry them out.

One of our colleagues, Chris Bonner, frequently uses a statement to explain to executives the importance of execution. It is appropriate to this chapter. Chris has stated, "A great strategy with average execution will always lose to an average strategy with great execution." This road map does not attempt to answer every question that must be answered. It is not intended to be the greatest strategy. It does not provide all of the detail your organization may need to carry out the change. It does define the major activities that upper managers and change agents must include in their change process if the desire is to achieve sustainable results. Whether your organization has begun a change process, or is contemplating one, this road map can be useful. For those who are beginning their journey, it provides a set of sequential activities that need to be implemented to achieve sustainable change. For those who may have begun their change, it can be a useful tool to identify gaps in your current plan or road map.

It is also important to note that we have not tried to identify all of the possible change or transformation approaches and include them. We have been focusing on achieving Six Sigma breakthroughs and beyond. Therefore, we have provided a road map for this type of change. However, the road map is appropriate for other types of changes as well.

Dr. Juran, during his 60 years of focus on implementing breakthroughs in organizations, has found that all organizational change must be integrated into an organization slowly to give the organization time to adapt to the change. His experience and ours has demonstrated repeatedly that organizational change is very similar to an antibiotic attacking a virus. Consider this scenario. A patient takes a drug daily, slowly building up an antibody to attack a virus (the old culture). As the virus (old culture) invites this drug or antibody into its system, the drug begins to kill the virus and a new health (culture) takes over, often leaving the patient healthier and free of this virus. It is our experience that the best change processes must invade a culture slowly, through a pilot effort, to begin to attack the culture. This in turn enables that same culture to accept the change. In other words, the change was tested on the culture and it worked. At that point, an organization could expand the changes to other parts of the organization with greater success and less resistance. Our road map includes a pilot phase to test out the intended change on the organization. This prepare phase can be used as a testing ground for any change being contemplated in the organization.

Before we describe the road map for change, let us review what we are changing from and to. A typical organization may look like this (and wants to change):

- High operating costs and lower than expected profit
- Productivity comes first over customers and quality, often sacrificing quality for schedules
- Slow to respond to complaints or needed corrective action

- Lack of creativity and initiative in new process or product and service designs
- Lack of employee–management trust leading to turnover or possible unionization
- Little collaboration or employee participation
- Quality is someone else's job

This organization would rather look like this (best in class):

- Greater profitability through leaner processes and greater productivity
- A customer-focused staff responding quickly to complaints
- Continuous quality improvement of all processes
- Employees empowered and in a state of self-control to maintain performance
- Flat, flexible organizational structures
- Quality is everyone's job
- Vision-driven leadership
- Values: high morals, ethics, teamwork, involvement, and risk taking

An organization that is not meeting executive, shareholder, and customer satisfaction is an organization that needs to change. The resources, energy, and desire to change must determine the level of change needed. It is difficult to quantify change, but looking at Figure 12.1, one can get an idea that transformational change or radical change is different from incremental change. A breakthrough can mean different things to different people. This graphic lists tools that are often associated with each of the different levels of change. The road map for change is focusing on the radical changes required. If your organization is not looking for radical change, the scope of your change process should be modified accordingly.

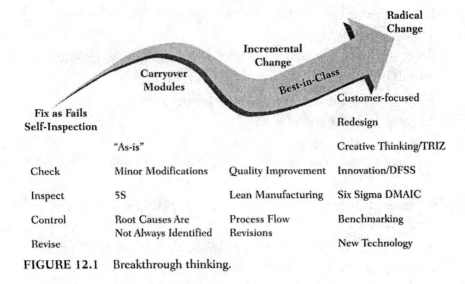

FIGURE 12.1   Breakthrough thinking.

THE ROAD MAP FOR CHANGE

The road map for change includes five phases. Each phase is independent of each other, but the beginning and end of each phase is not as clear as it may seem. Each organization reacts differently to changes taking place. This can lead to one business unit in an organization remaining in one phase longer than another business unit does. These phases once again are a managerial guide to change, not a prescription.

The five phases of our road map are shown in Figure 12.2.

The road begins with the decide phase. This phase begins with someone on the executive team deciding something must be done or the organization will not meet its shareholder expectations, will not meeting its plan, etc. It ends with a clear plan for change (Figure 12.3).

In the decide phase, the organization will need to create new information or better information than it may have had about itself. This information can come from a number of reviews or assessments it can make. Our experience shows that the more new information an organization has, the better its planning for change will be. There are a number of areas that should be reviewed. Here are some of the important ones:

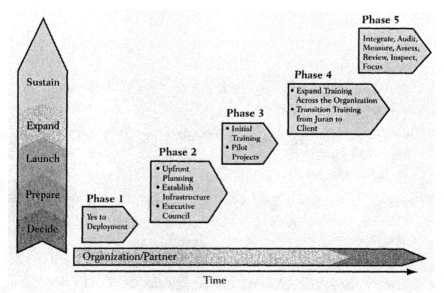

FIGURE 12.2   Using the road map maximizes the probability of success and avoids the "flavor of the month" syndrome.

- For your customers:

  Conduct a customer loyalty assessment to determine what they like or dislike about your products and services

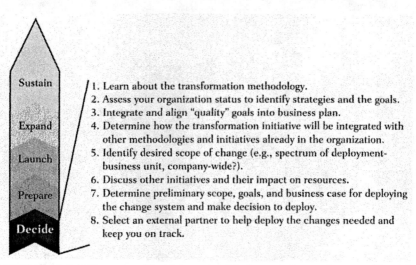

FIGURE 12.3   Phase one (decide) starts with the questions—do we and how do we pursue the transformation?

- For your culture:

  Identify the areas of strength and uncover possible problems in the organization's performance

  Understand their attitude toward the proposed changes

- Key business processes.

  Understand the key business processes and how they will be impacted by the changes

- Determine the business case for change:

  Conduct a cost analysis of poor performing processes to determine the financial impact of theses costs on the bottom line

- Conduct a world-class business quality system review of all of the business units to understand the level of improvement needed in each unit

A comprehensive review of the organization prior to launch is essential for success. A typical review we would recommend to all organizations embarking on a Six Sigma transformation initiative is shown in Figure 12.4.

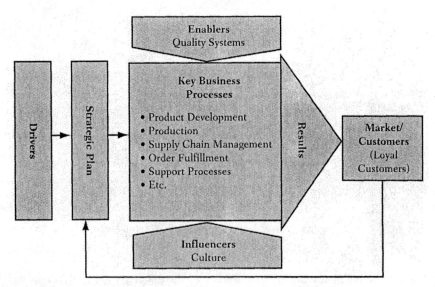

FIGURE 12.4   Conduct a comprehensive organizational review.

From these assessments and reviews, the executive team now has qualitative and quantitative information to define the implementation plan for its organization. This plan must include:

- The infrastructure needed to steer the changes
- The methodology and tools that will be used throughout the implementation
- The goals and objectives of the effort
- The detailed plan for achieving results

The second phase is the prepare phase. In this phase, the executive team begins to prepare for the upcoming changes that will take place. It focuses on developing a pilot effort to try out the change in a few business units before carrying it out in the total organization. This phase begins by deploying the plan created in phase one, and it ends after a successful launch of the pilot projects (Figure 12.5).

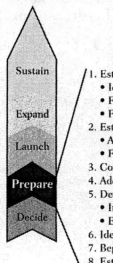

1. Establish infrastructure:
    - Identify the change initiative leader
    - Form the executive council
    - Form other steering committee(s)
2. Establish the initiative goals:
    - Assure they are assigned to company goals
    - Finalize business case
3. Conduct HR planning.
4. Address finance issues.
5. Develop and begin implementation of the communications plan:
    - Internal
    - External
6. Identify and select pilot projects for training and beyond.
7. Begin developing a balanced business scorecard.
8. Establish appropriate metrics and initiative tracking.
9. Finalize deployment plan.

FIGURE 12.5   Phase two (prepare) builds on the reviews and sets up the organization for success.

From here, the organization begins to identify the potential improvement projects that must be carried out to meet the desired goals established in the decide phase. In this phase, the organization launches the pilot projects, reviews the pilots' progress, and enables success. Upon completion of the pilot projects, the executives evaluate what has worked and what has not. They then either abandon their efforts, or in most cases, make necessary changes to the plan and expand it throughout the organization (Figure 12.6).

Expansion can take months or years, depending on the size of the organization. A company of 500 employees will require less time to deploy across the company than an organization of 50,000. The expand phase may take three to five years. It is important to note that positive financial results will occur long before cultural changes takes place. Staying in the expand phase is not a bad thing. An organization must continue to implement business unit by business unit, company by company, until all of the organization has had enough time to implement the desired changes (Figure 12.7).

The final phase is the sustain phase. This is the phase when the organization has a fully integrated operation. All improvement and Six Sigma goals are aligned with the strategy of the

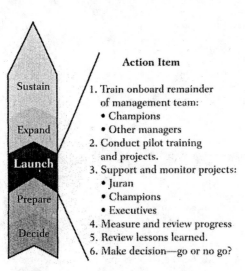

Action Item

1. Train onboard remainder of management team:
   - Champions
   - Other managers
2. Conduct pilot training and projects.
3. Support and monitor projects:
   - Juran
   - Champions
   - Executives
4. Measure and review progress
5. Review lessons learned.
6. Make decision—go or no go?

How Partners Can Help

1. Executive and champion training, mentoring, and coaching.
2. Employee training, mentoring, and coaching.
3. Project support.
4. Establishment of key metrics.
5. Analysis of progress.
6. Input into decision process.

FIGURE 12.6   Phase three (launch).

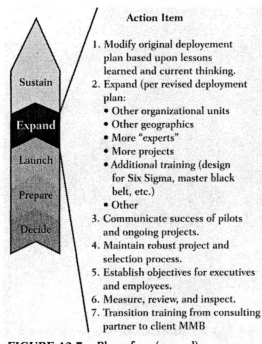

Action Item	How Partners Can Help
1. Modify original deployement plan based upon lessons learned and current thinking.	1. Executive and champion training, mentoring, and coaching.
2. Expand (per revised deployment plan: • Other organizational units • Other geographics • More "experts" • More projects • Additional training (design for Six Sigma, master black belt, etc.) • Other	2. Employee training, mentoring, and coaching. 3. Transition planning and implementation. 4. Train the trainers. 5. Licensed materials.
3. Communicate success of pilots and ongoing projects.	
4. Maintain robust project and selection process.	
5. Establish objectives for executives and employees.	
6. Measure, review, and inspect.	
7. Transition training from consulting partner to client MMB	

FIGURE 12.7   Phase four (expand).

organization. Key business processes are defined and well managed, and a process owner is assigned to manage them. Employee performance reviews and compensation are in line with the changes required. Reward happens for those complying with the change. The executives and business unit heads conduct regular reviews and audits of the change process. This may result in a discussion or even a change in the strategy of the organization (Figure 12.8).

The organization may have learned more about its capabilities and more about its customers. This may lead to a change in strategy. The sustain phase also endures as long as the organization is meeting its strategy and financial results. Deviations from results expected, possibly due to macro-economic events occurring outside of the organization, require a review of the score card to determine what has changed. When this is determined, the organization makes the changes, continues, and sustains itself at the current level (Figure 12.9).

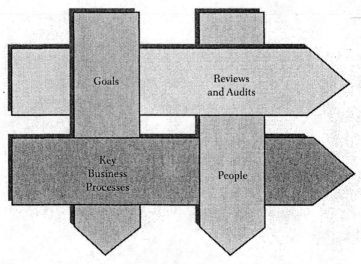

FIGURE 12.8 There are four dimensions for successful integration of the intended change into the life of an organization.

SUCCESS FACTORS ON DEPLOYING THIS ROAD MAP

As you begin your journey down this road, it is important to note that there are many lessons learned from organizations that have led a change process and initially failed. These failures can be avoided if planned for. Here are some of them:

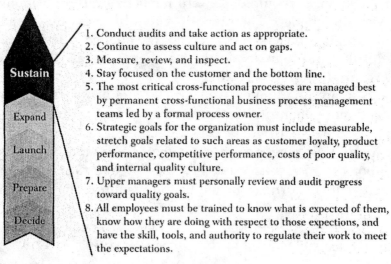

1. Conduct audits and take action as appropriate.
2. Continue to assess culture and act on gaps.
3. Measure, review, and inspect.
4. Stay focused on the customer and the bottom line.
5. The most critical cross-functional processes are managed best by permanent cross-functional business process management teams led by a formal process owner.
6. Strategic goals for the organization must include measurable, stretch goals related to such areas as customer loyalty, product performance, competitive performance, costs of poor quality, and internal quality culture.
7. Upper managers must personally review and audit progress toward quality goals.
8. All employees must be trained to know what is expected of them, know how they are doing with respect to those expections, and have the skill, tools, and authority to regulate their work to meet the expectations.

FIGURE 12.9 Phase five (sustain).

- Internal experts (Six Sigma black belts) become drivers who propel their company to best-in-class.
- Extensive training in tools and techniques for all employees assures learning has taken place and they can use the tools learned.
- Systematic application and deployment through proven methodologies like the Six Sigma DMAIC and DFSS are necessary to create a common language.
- Focusing improvements on the customer first will enable cost reduction and delighted customers, which will enable breakthrough bottom-line results.
- Significant increase in customer satisfaction only happens when you improve on the processes and services that influence them.
- No company has ever achieved a successful implementation without the leadership and commitment of the executive team—they control the resources and the culture.

With this road map and the lessons learned in hand, all organizations should be able to achieve sustainable results well into the future. If more organizations get on board with positive, customer-focused change initiatives, we will be able, as a nation or many nations, to compete with any of the low cost producing nations and competitors. Instead of China or Korea becoming global leaders and negatively impacting our societies, the United States and Europe can maintain their competitiveness and survive the future.

A new revolution in quality is needed. It can begin with your organization or it can begin with your competitor. It is your choice. Many years ago, Dr. Juran ended a very famous videotape learning series on quality improvement with the words "long live the revolution." He was of course referring to the revolution that was taking place in the Western world to win back our industries so beaten up by the Japanese quality. We did it then, and we will do it again against China, Korea, and any tough competitor. In a slight revision of a Joseph Juran quote: in the words of Dr. Juran (and the authors) "Long live the 'next' revolution."

REFERENCES

CHAPTER 2 REFERENCES

De Feo JA. "The Tip of the Iceberg." In: *Quality Progress*. American Society for Quality; May 2001.

Juran JM. *Juran on Leadership for Quality*. New York: The Free Press; 1989.

Juran JM, Godfrey AB (editors). *Juran's Quality Handbook*. New York: McGraw-Hill; 1999.

Juran JM. *Managerial Breakthrough*. New York: McGraw-Hill; 1995.

CHAPTER 3 REFERENCES

Design for World Class Quality. Wilton, CT: Juran Institute, Inc.; 1995.

Innovative Design for Six Sigma: Black Belt Training (Workbook). Wilton, CT: Juran Institute, Inc.; 2003.

Juran JM. *Juran on Leadership for Quality*. New York: The Free Press; 1989.

Juran JM. *Juran on Quality by Design*. New York: The Free Press; 1992.

CHAPTER 4 REFERENCES

Burgam PM. "Application: Reducing Foundry Waste." *Manufacturing Engineering*, March 1985.

Bylinsky G. "How Companies Spy on Employees." *Fortune*. November 1991: 131–140.

Carr WE. "Modified Control Limits." *Quality Progress*. January 1989: 44–48.

Deming WE. *Out of the Crisis.* Cambridge, AM: MIT Center for Advanced Engineering Study; 1986.

Deming WE. *Elementary Principles of the Statistical Control of Quality.* Tokyo: Nippon Kagaku Gijutsu Renmei (Japanese Union of Scientists and Engineers); 1950.

Duyck TO. "Product Control Through Process Control and Process Design." 1989 *ASQC Quality Congress Transactions.* 1989: 676–681.

Gass KC. "Getting the Most Out of Procedures." *Quality Engineering.* June 1993.

Goble J. "A Systematic Approach to Implementing SPC." 1987 *ASQC Quality Congress Transactions.* 1987: 154–164.

Juran JM. *Juran on Quality by Design.* New York: The Free Press; 1992.

Juran JM (editor). *Juran's Quality Control Handbook.* New York: McGraw-Hill; 1988.

Juran JM. *Managerial Breakthrough.* New York: McGraw-Hill; 1964.

Juran JM, Godfrey AB (editors). *Juran's Quality Handbook*, fifth edition. New York: McGraw-Hill; 1999.

Koura K. "Deming Cycle to Management Cycle." Societas Qualitas, Japanese Union of Scientists and Engineers, Tokyo, May–June 1991.

Lenehan M. "The Quality of the Instrument, Building Steinway Grand Piano K 2571." *The Atlantic Monthly.* August 1981: 32–58.

CHAPTER 5 REFERENCES

De Feo JA. "The tip of the iceberg." In: Quality Progress. *American Society for Quality*, May 2001.

Gryna FM. "Quality and costs." In: Juran JM, Godfrey AB (editors). *Juran's Quality Handbook.* New York: McGraw-Hill; 1999.

Haines SG. *The Manager's Pocket Guide to Systems Thinking and Learning.* Amherst, MA: HRD Press; 1998.

Juran JM. *Managerial Breakthrough.* New York: McGraw-Hill; 1995.

Kanter RM, Kao J, Wiersewa F. *Innovation.* New York: Harper Collins Publishers, Inc.; 1997.

Katz D, Kahn RL. *The Social Psychology of Organization.* New York: John Wiley & Sons, Inc.; 1966.

Lippitt GL. *Organization Renewal* (2nd edition). Englewood Cliffs, NJ: Prentice-Hall, Inc. 1982.

CHAPTER 6 REFERENCES

Ambrose SE. *The Supreme Commander*. Jackson, MS: University of Mississippi Press; 1999.

De Feo JA, Janssen JA. "The values of strategic deployment." Measuring Business Excellence. *MCB UP Limited*, Vol. 6, No. 1, 2002.

Harper SC. *The Forward-Focused Organization*. New York: AMA-COM; 2001.

Juran JM. *Juran on Leadership for Quality*. New York: The Free Press; 1989.

Kanter RM. *The Change Masters*. New York: Simon & Schuster; 1983.

Kaplan RS, Norton DP. *The Balanced Scorecard*. Boston: Harvard Business School Press; 1996.

Kotter JP. *Leading Change*. Boston: Harvard Business School Press; 1996.

Pietersen WG. *Reinventing Strategy*. New York: John Wiley & Sons, Inc.; 2002.

Senge PM. *The Fifth Discipline*. New York: Doubleday Currency; 1990.

Smith DK. *Taking Charge of Change*. Reading, MA: Addison Wesley; 1996.

Tjosuold DW, Tjosuold MM. *Leading the Team Organization*. New York: Lexington Books; 1991.

CHAPTER 7 REFERENCES

Basu R, Wright JN. *Quality Beyond Six Sigma*. Oxford: Butterworth Heinemann; 2003.

Brue G. *Six Sigma for Managers*. New York: McGraw-Hill; 2002.

Collins J. *Good To Great*. New York: Harper Business; 2001.

Day JD. *Organizing for Growth*. McKinsey; Quarterly, 2001.

Day JD, Mang PY, Richter A, Roberts J. *The Innovative Organization*. McKinsey; Quarterly, 2001.

de Geus A. *The Living Company Habits for Survival in a Turbulent Business Environment*. Boston: Harvard Business School Press; 1997.

Dichter SF, Gagnon C, Alexander A. *Leading Organizational Transformations*. McKinsey; Quarterly, 1993.

Ehrlich BH. *Transactional Six Sigma and Lean Servicing*. Boca Raton, FL: St.Lucie Press; 2002.

George ML. *Lean Six Sigma*. New York: McGraw-Hill; 2002.

Häcki R, Lighton J. *The Future of the Networked Company*. McKinsey; Quarterly, 2001.

Juran JM, Godfrey BA. *Juran's Quality Handbookm* 5th edition. Kotter JP. Leading Change. Boston: Harvard Business School Press; 1996.

Nadler DA. *Champions of Change: How CEOs and Their Companies Are Mastering the Skills of Radical Change*. San Francisco: Jossey Bass; 1998.

Singerm M. *Beyond the Unbundled Corporation*. McKinsey; Quarterly, 2001.

Slater R. *Jack Welch and the GE Way*. New York: McGraw-Hill; 1999.

Smith D, Blakeslee J, Koonce R. *Strategic Six Sigma Best Practice from the Executive Suite*. Hoboken, NJ: John Wiley & Sons, Inc.; 2002.

Ulrich D, Zenger J, Smallwood N. *Results Based Leadership*. Boston: Harvard Business School Press; 1999.

CHAPTER 8 REFERENCES

Six Sigma DMAIC Black Belt Training. Workbook. Wilton, CT: Juran Institute, Inc.; 2003.

Juran JM. *Juran on Leadership for Quality*. New York: The Free Press; 1989.

Juran JM. *Managerial Breakthrough*. New York: McGraw-Hill; 1995.

Juran JM, Godfrey BA. *Juran's Quality Handbook*. 5th ed. New York: McGraw-Hill; 1999.

CHAPTER 9 REFERENCES

Juran JM. *Managerial Breakthrough*. New York: McGraw-Hill; 1995.

Juran JM. "Improving the Relationship between Staff and Line—An Assist from the Anthropologists," Article, 1956. Listed in Juran Institute's website: www.Juran.com. Reprinted as a "Mentor Book" by the New American Library; 1955.

Juran JM. "Cultural Patterns and Quality Control," paper presented at the Thirteenth Annual Quality Control Clinic, Rochester, NY, February 19, 1957.

Katz D, Kahn RL. *The Social Psychology of Organizations*. New York: John Wiley & Sons, Inc.;1966.

Meade M, ed. *Cultural Patterns and Technical Change*. Paris: United Nations Educational Scientific and Cultural Organization; 1953.

CHAPTER 10 REFERENCES

Haeckel SH. *Adaptive Enterprise—Creating and Leading Sense-and-Respond Organizations*. Boston, MA: Harvard Business School Press; 1999.

Pall GA. *The Process-Centered Enterprise—The Power of Commitments*. Boca Raton, FL: St. Lucie Press; 2000.

Redman TC. *Data Quality for the Information Age*. Boston, MA: Artech House; 1996.

Redman TC. *Data Quality Management and Technology*. New York: Bantam Books; 1992.

CHAPTER 11 REFERENCES

Branco GJ, Willoughby RS. *Extending Quality Improvement to Suppliers*. Fourth Annual Conference on Quality Improvement, IMPRO 86. Wilton, CT: Juran Institute, Inc.; 1987.

Brunetti W. "Policy Deployment—A Corporate Roadmap." Proceedings of Fourth Annual Conference on Quality Improvement (IMPRO 86). Wilton, CT: Juran Institute, Inc.; 1987:20–29.

Delaplane GW. *Integrating Quality Into Strategic Planning* (IMPRO 87). Wilton, CT: Juran Institute, Inc.; 1987:21–29.

Godfrey AB, Chua RC. Integrating Productivity and Quality with Strategic Business Performance. *Productivity Digest*. July 1997.

Gryna FM. The quality director of the '90s. *Quality Progress*. April 1991.

Gryna FM. The quality director of the '90s: assisting upper management with strategic quality management. *Quality Progress*. May 1991.

Ikezawa T, Kondo Y, Harada A, Yoneyama T. Features of Companywide Quality Control in Japan, Report of 44th QC Symposium. International Conference on Quality Control (ICQC). Tokyo, Japan: Japanese Union of Scientists and Engineers; 1987.

Ishikawa K. The quality control audit. *Quality Progress.* January 1987:39–41.

Juran JM. *Managerial Breakthrough.* New York: McGraw-Hill; 1964.

Juran JM. ed-in-chief. *Juran's Quality Control Handbook.* 4th ed. New York: McGraw-Hill Book Company; 1988.

Juran JM. *Juran on Planning for Quality.* New York: The Free Press, A Division of Macmillan, Inc.; 1988.

Kegarise RJ, Miller GD. An Alcoa-Kodak Joint Team. Proceedings of Annual Conference on Quality Improvement (IMPRO 85). Wilton, CT: Juran Institute Inc.;1986:29–34.

Kondo Y. In Juran's *Quality Control Handbook.* 4th edition New York: McGraw-Hill Book Organization; 1988. Kondo provides a detailed discussion of quality audits by Japanese top managements, including the president's audit. See Section 35F, Quality in Japan, under Internal QC Audit by Top Management.

Manshi KF. *The Importance of a Clear Corporate Vision in Policy Deployment.* IMPRO 91. Wilton, CT: Juran Institute, Inc.; 1991.

McGrath JH. Successful Institutionalized Improvement in Manufacturing Areas. Third Annual Conference on Quality Improvement, IMPRO 85. Wilton, CT: Juran Institute, Inc.; 1986.

Onnias A. *The Quality Blue Book.* Proceedings of Third Annual Conference on Quality Improvement (IMPRO 85). Wilton, CT: Juran Institute, Inc.; 1986:127–131.

Pisano DJ Jr. Replanning the Product Development Process. Proceedings of Fourth Annual Conference on Quality Improvement (IMPRO 86). Wilton, CT: Juran Institute, Inc.; 1987:260–264

Shimoyamada K. The President's audit: QC audits at Komatsu. *Quality Progress.* January 1987: 44–49.

Weigel PJ. Applying Policy Deployment Below the Corporate Level, IMPRO 90. Wilton, CT: Juran Institute, Inc.; 1990.

Wolf JD. Quality Improvement: The Continuing Operational Phase. Second Annual Conference on Quality Improvement, IMPRO 84. Wilton, CT: Juran Institute, Inc.; 1985.

INDEX

Note: Boldface numbers indicate illustrations.

ABOUT THE AUTHORS

Joseph A. De Feo, President and Chief Executive Officer of Juran Institute, is recognized for his training and consulting expertise in sustaining breakthrough improvements within many different organizations worldwide. He has been instrumental in progressing the subject of Breakthrough Improvement by focusing on the needs of today's contemporary and demanding industries. He is considered a knowledge expert on the Management of Quality, Business Process Management, Six Sigma Deployment, Strategic Quality Planning, and supportive methodologies. He is a regular speaker at national conferences and his counsel has been published widely in national and international publications.

William Barnard, PhD. is a Senior Vice President at Juran Institute. Bill advises and assists management in industries such as health care, universities, and government ministries, emphasizing how to implement managing for quality. He has represented Juran Institute in North America, Europe, Asia, Africa, and the Pacific and has originated the Institute's Facilitator Training courses.

Juran Institute, Inc. provides training and consulting services to help organizations worldwide achieve sustainable breakthrough improvements. Dr. Joseph M. Juran, a pioneer in the quality revolution who developed many of the techniques and tools on which the Six Sigma methodology is based, founded the Institute in 1979. Apart from Six Sigma, the Institute's areas of expertise include breakthrough improvement, strategic deployment, benchmarking and lean manufacturing. Juran Institute is headquartered in Wilton, Connecticut, and has European offices in Amsterdam and Madrid. For more information, visit www.juran.com